Monitoring Environmental Contaminants

Environmental Contaminants

Monitoring Environmental Contaminants

Series Editors

Alistair B.A. Boxall

Professor, Environmental Department, University of York, United Kingdom

Rai S. Kookana

Chief Research Scientist, CSIRO Land and Water, Australia

Edited by

Steven Johnson

Head of Intelligent Systems and Nano-science Research Group, University of York, United Kingdom

ELSEVIER

Elsevier
Radarweg 29, PO Box 211, 1000 AE Amsterdam, Netherlands
The Boulevard, Langford Lane, Kidlington, Oxford OX5 1GB, United Kingdom
50 Hampshire Street, 5th Floor, Cambridge, MA 02139, United States

Notices

Knowledge and best practice in this field are constantly changing. As new research and experience broaden our understanding, changes in research methods, professional practices, or medical treatment may become necessary.

Practitioners and researchers must always rely on their own experience and knowledge in evaluating and using any information, methods, compounds, or experiments described herein. In using such information or methods they should be mindful of their own safety and the safety of others, including parties for whom they have a professional responsibility.

To the fullest extent of the law, neither the Publisher nor the authors, contributors, or editors, assume any liability for any injury and/or damage to persons or property as a matter of products liability, negligence or otherwise, or from any use or operation of any methods, products, instructions, or ideas contained in the material herein.

Library of Congress Cataloging-in-Publication Data
A catalog record for this book is available from the Library of Congress

British Library Cataloguing-in-Publication Data
A catalogue record for this book is available from the British Library

ISBN: 978-0-444-64335-3

Publisher: Candice Janco
Acquisitions Editor: Marisa LaFleur
Editorial Project Manager: Joshua Mearns
Production Project Manager: Bharatwaj Varatharajan
Cover Designer: Greg Harris

Typeset by SPi Global, India

Contents

[†] Deceased.

Contributors

Bidroha Basu School of Engineering, University College Dublin, Dublin, Ireland

Alexander L. Bond Bird Group, Department of Life Sciences, The Natural History Museum, Tring, United Kingdom

Jacqueline S. Chaplow UK Centre for Ecology & Hydrology, Lancaster, United Kingdom

Geert Cornelis Department of Soil and Environment, Swedish University of Agricultural Sciences, Uppsala, Sweden

Dermot Diamond Insight Centre for Data Analytics, National Centre for Sensor Research, School of Chemical Sciences, Dublin City University, Dublin, Ireland

Andrew Donohoe Insight Centre for Data Analytics, National Centre for Sensor Research, School of Chemical Sciences, Dublin City University, Dublin, Ireland

Julian A. Gallego-Urrea Department of Marine Sciences, University of Gothenburg, Fiskebäckskil, Sweden

Salem Gharbia Department of Civil Engineering & Construction, IT SLigo, Sligo, Ireland

Andreas Gondikas Faculty of Geology and Geo-environment, National and Kapodistrian University of Athens, Athens, Greece

Panagiota Katsikouli DTU COMPUTE Department of Applied Mathematics and Computer Science, Technical University of Denmark, Lyngby, Denmark

Jan Koschorreck Federal Environment Agency, Berlin, Germany

Lee Markham Connected Places Catapult, London, United Kingdom

Karin Mattsson Department of Marine Sciences, University of Gothenburg, Fiskebäckskil, Sweden

Margaret McCaul Insight Centre for Data Analytics, National Centre for Sensor Research, School of Chemical Sciences, Dublin City University, Dublin, Ireland

Manuel Montaño Huxley College of the Environment, Western Washington University, Bellingham, WA, United States

Francesco Pilla School of Engineering; Spatial Dynamics Lab, College of Engineering, University College Dublin, Dublin, Ireland

Heinz Rüdel Fraunhofer Institute for Molecular Biology and Applied Ecology IME, Schmallenberg, Germany

Katinka Schaaf Connected Places Catapult, London, United Kingdom

Richard F. Shore[†] UK Centre for Ecology & Hydrology, Lancaster, United Kingdom

Jani Tuoriniemi Department of Soil and Environment, Swedish University of Agricultural Sciences, Uppsala, Sweden

Stephan Wagner Department of Analytical Chemistry, Helmholtz-Centre for Environmental Research—UFZ, Leipzig, Germany

John Wilkinson Environment and Geography Department, University of York, York, United Kingdom

[†] Deceased.

High-performance liquid chromatography-tandem mass spectrometry for analysis of aquatic contaminants: A high-level introduction to the technique

1

John Wilkinson
Environment and Geography Department, University of York, York, United Kingdom

1 Introduction

For many years, the study of trace organic contaminants in water was largely limited by the sensitivity of analytical techniques. With the introduction of analytical methods such as high-performance liquid chromatography (HPLC) coupled with tandem mass spectrometry (MS/MS), detection limits have been improved to an extent where trace-level analysis (ng/L or below) can now be undertaken for the occurrence and fate of organic contaminants in the environment. By the first decade of the 21st century, techniques such as HPLC-MS/MS and solid phase extraction (SPE) were commonly and are still successfully used to quantify the levels of thermally unstable, nonvolatile, or polar organic contaminants at environmentally relevant (low ng/L) concentrations (Kolpin et al., 2002; Kim et al., 2009; Furlong et al., 2014; Wilkinson et al., 2019). The ecotoxicity of some of these contaminants in the aquatic environment (e.g., Wilkinson et al., 2016; Neilson, 2017) has precipitated a multitude of studies on academic, industrial, and governmental levels, many of which were underpinned by HPLC coupled to MS. The technique has emerged as a crucial tool for the study of aquatic and environmental science.

Here, a high-level overview of HPLC-MS/MS methodology is presented for the analysis of organic contaminants in water. Topics of discussion include sample collection, preparation/pretreatment protocols, SPE, HPLC theory and methodology, matrix effects, MS theory and methodology, brief discussion of other types of MS used with HPLC, quantitation, and quality assurance/control. The text aims to present the technique of HPLC-MS/MS for the analysis of organic contaminants in water at an entry to mid-level.

Monitoring Environmental Contaminants. https://doi.org/10.1016/B978-0-444-64335-3.00004-9

2 Sample collection

Methodologies employed for water sample collection largely depend on the analytical pretreatment required and vary from a small volume of (1–10 mL) grab sampling to time-series composite sampling. Passive sampling techniques are particularly useful in certain instances, however, and will not be reviewed here. These protocols are well presented elsewhere (e.g., Vrana et al., 2005; Roll and Halden, 2016).

Grab sampling (Fig. 1A) is the practice of collecting a sample or measurement at one time and in a single place. In the context of water sampling, this may include replicates collected at the same place and at the same time. This sampling technique is commonly used in environmental monitoring campaigns due to the relative ease and low cost compared to the use of passive or composite samplers. In a well-mixed aquatic system (e.g., a small river downstream from a waterfall), grab sampling can be a specific and well-representative "snap-shot" measure of contamination. However, the technique has well-documented drawbacks. Notably, grab sampling inherently represents only a "snap-shot" view of contamination in and at a specific place in time, which may not be representative of the situation mere moments later. For example, in a large river, grab samples taken from either bank and in the centroid flow are likely to represent differing environmental and physical profiles. Shallower flows along the banks may move at a rate different from the centroid flow and may be at a higher relative temperature than the deeper water. Similarly, flows along either bank may be differentially affected by anthropogenic discharges such as agricultural runoff or point-source wastewater treatment plant effluents. Hence, care must be exercised when drawing conclusions based on monitoring data derived from grab sampling techniques.

Despite the increased cost and complexity, composite/sequential samplers offer key advantages over traditional grab sampling techniques. As opposed to the "snap-shot" view of grab sampling, composite samples may provide a more

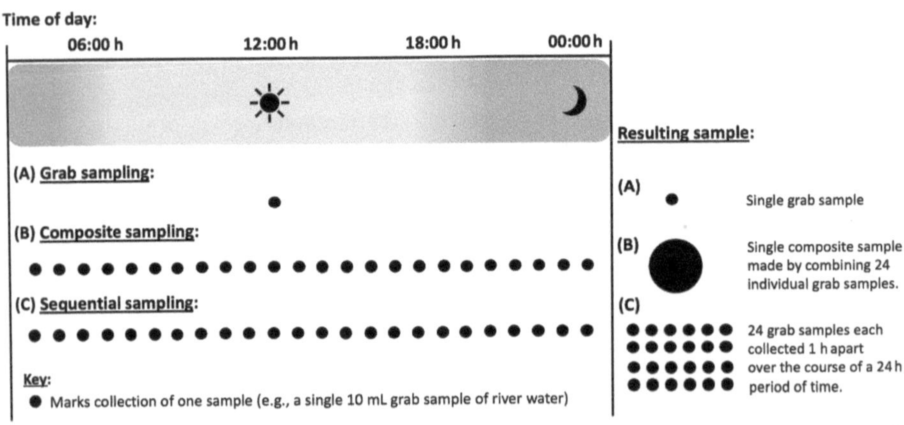

Fig. 1 Comparison of the type, frequency, and resulting sample obtained via an example of traditional grab sampling (A), composite sampling (B), and sequential sampling (C) over a 24-h period assuming hourly composite and sequential sampler collection.

representative insight into the extent of a given variable over time. Such instruments are capable of collecting multiple grab samples at a given location and frequency. The frequency of sample collection can depend on variables such as time (e.g., hourly) or flow (e.g., after every 1000 L). Typically, composite samplers will pool all of the grab samples taken over the desired time period, resulting in one large sample (Fig. 1B). For example, the composite sampler may be programmed to pump 10 mL of water from a given point along a river into a holding tank once every hour for 24 h, resulting in a final 240 mL sample (24 $\times$ 10 mL samples). This composite of the hourly grab samples would therefore represent an average environmental exposure at that single point over the sampling period.

Similarly, sequential sampling may also be programmed to take grab samples at a specific point, frequency, and duration. However, these samplers differ from composites as the individual grab samples remain separate (i.e., they are not combined into a single composite), thereby allowing for the characterization of exposure variability throughout a period of time (Fig. 1C). Although sequential samples may aid in the elucidation of temporal or diurnal effects on the collected medium (e.g., sunlight, temperature, river flow, and time of day on river contaminant concentrations), the ultimate increased number of resulting samples may concomitantly increase the complexity, time, and cost of analysis. Hence, optimization of the sampling protocol most applicable to specific scenarios should be undertaken to balance analytical rigor and sample representation.

3 Water sample preparation and the effect of the matrix

The aim of sample pretreatment is to deliver the target analytes in a solution and at a concentration appropriate for the selected analytical method (e.g., HPLC-MS/MS). For the analysis of organic chemicals in water, sample preparation protocols are optimized based on.

(a) sensitivity required to answer the research questions,
(b) physiochemical characteristics of the target analytes,
(c) type of aqueous sample (e.g., saltwater, groundwater, wastewater, and sewage),
(d) sensitivity of the analytical equipment used for analyte quantification,
(e) equipment requirements/maintenance, and.
(f) time and financial considerations.

The most common sample preparation techniques include filtration, SPE, and dilution (Table 1). During sample pretreatment, either the target analytes are separated from interfering matrix materials, which may affect sample analysis, or these materials are simply diluted in the injected samples.

3.1 Sample matrix and effects on analysis

The so-called interfering matrix materials, which are capable of affecting instrumentation, sample preservation, and analysis of the target analytes, are present in most environmental samples. All of these potential effects may be significant negative

Table 1 Common preparation/pretreatment techniques for water samples prior to HPLC-MS/MS analysis with notable advantages and disadvantages of each technique.

Technique	Aim(s)	Key advantages	Potential disadvantages
Filtration	To remove particulate material from water (typical filter pore sizes: 0.45–0.7 μm)	- Considered a gold standard of sample preparation and a prerequisite to most other cleanup protocols - Removes fine materials in water, which may otherwise have damaged or reduced the chromatography column lifetime - Removes bacteria, which may degrade target chemicals in the samples - Cheap and easy	- Some organic contaminants may slightly adsorb onto the filter material itself - Potential for the introduction of chemicals originating from the filters themselves or the materials involved in the filtration protocol (e.g., plastic syringes and cross-contamination using a Buchner funnel) - Nonselective
Dilution	To decrease the concentration of interfering matrix materials in the water sample	- Reduces the concentration of matrix materials (e.g., humic acids, and salts), which may interfere with the MS response, dirty the column/instruments or coelute with analytes - Can reduce the concentration of target analytes to within the calibrated range in highlycontaminated samples (e.g., wastewater) - Cheap, fast and easy to automate	- Dilutes the target analytes and increases the limit of detection - Does not remove interfering matrix materials from the water but only dilutes them - Interfering matrix materials may cause damage to the chromatography column and/or mass spectrometer over time, if injected directly - Nonselective

Table 1 Continued

Technique	Aim(s)	Key advantages	Potential disadvantages
Solid phase extraction	- To remove interfering matrix materials from the water sample - To concentrate the analytes prior to analysis	- Predominately removes most of the interfering matrix materials that may interfere with the MS response, dirty the column/ instruments, or coelute with analytes - Concentrates the analytes in the sample, which increases sensitivity and lowers limits of detection. - Enables solvent exchange (e.g., analytes can be exchanged into the mobile phase, which may improve the chromatography) - Selective cleanup	- Time intensive and relatively expensive - Requires careful selection of appropriate sorbent material - Requires additional equipment - Potential loss of analytes if the protocol is not well optimized - Requires the use of extraction solvents

factors on accuracy, precision, variability, and the robustness of HPLC-MS/MS analytical protocol. In water, such materials include particulate materials, salts, humic acids, phospholipids, and microorganisms (which may degrade organic chemicals). These materials may accumulate in the mass spectrometer components, alter the MS response to the analytes or coelute during chromatography (Gosetti et al., 2010). For example, salts present in the injected samples may precipitate at high temperatures and accumulate in the preliminary stages of MS analysis, reducing instrument sensitivity over time.

Mass spectrometry response to the analytes may be affected by signal suppression or enhancement (Gosetti et al., 2010). While their mechanisms are not completely known, both arise out of competition between the targeted analytes and materials present in the matrix during the various stages of MS analysis (Gosetti et al., 2010). Signal suppression is marked by the loss of MS response for a given analyte and may result in artificially low limits of detection or false negatives. Suppression may arise, for example, via poorly volatile compounds in the matrix altering analyte droplet formation at the nebulizer, which in turn may reduce the amount of analyte in the gas phase ultimately reaching the MS detector (Annesley, 2003). Similarly, the enhancement of the MS response for an analyte can result in false positives or artificially high analyte concentrations. Enhancement may occur, for example, via the suppression of the MS response to an analyte's internal standard, thereby artificially increasing the slope of the resulting linear concentration response ratio during calibration.

3.2 Filtration

Filtration involves the physical separation of any undissolved matter, such as suspended particulate materials, microorganisms, and suspended algae from the water sample. This technique is simple, fast, and commonly considered a gold standard of sample pretreatment and a prerequisite to any additional preparation, including SPE. Furthermore, recent protocols have suggested immediate in-field filtration of small (<25 mL) sample volumes, eliminating any need for lab-based vacuum filtration methodologies (e.g., Furlong et al., 2014; Wilkinson et al., 2019). Filter media for water filtration consists of disks or sheets that comprise cellulose acetate or a glass membrane with pore sizes ranging from 0.45 to 0.7 μm. Typically, samples are vacuum filtered using a Buchner funnel in the lab or may be filtered by using a syringe when large volumes of the sample are not required. When using a particularly sensitive analytical equipment, filtration may be a sufficient pretreatment technique for samples without significant interfering matrix materials (e.g., for ground or surface water analysis). When a high degree of analytical sensitivity is needed or the sample is of a particularly challenging matrix (e.g., wastewater or sewage), additional sample pretreatment is likely needed beyond filtration.

3.3 Dilution

A simple, cheap, and efficient way to reduce the effect of interfering matrix materials is via sample dilution with pure (e.g., LCMS-grade reagent) water after filtration. While dilution does not remove interfering materials, it does reduce their concentrations. However, target analytes in the sample will also be diluted, thereby reducing sensitivity and ultimately detection limits. Sample dilution is often used where further sample cleanup via SPE is not necessary and the target analytes are of a sufficiently high concentration that their final concentration is within the calibration range. An example of dilution in environmental sample analysis is the direct injection of 50:50 treated wastewater/LCMS-grade reagent water (v:v) for the analysis of pharmaceutical residues following a 0.7-μm filtration step (e.g., Wilkinson et al., 2019). When using dilution as the predominate sample cleanup technique for HPLC, care must be taken to prevent long-term chromatography column or other instrument damage. As interfering materials are not removed via dilution, some of these constituents of the sample matrix (e.g., salts and ultrafine particulate material) will enter the HPLC and, potentially, the mass spectrometer. To sufficiently alleviate potential column damage, a guard column may be used upstream from the separation column. Similarly, to prevent undesirable material from entering the mass analyzer, the mobile phase flow may be diverted to waste (i.e., away from the mass spectrometer) for the first minute of the analytical run to allow for poorly retained elements (e.g., salts) to be removed.

3.4 Solid phase extraction

Solid phase extraction is a form of chromatography based on the principle of separating target analytes from a liquid matrix via their affinity to a solid sorbent material. The specific material used in SPE cartridges can be designed to select certain analytes whilst removing others and interfering matrix materials. The technique also allows for the enrichment of analytes in the injected sample, which ultimately increases the sensitivity of the employed method. Furthermore, by significantly reducing or eliminating the interfering matrix materials in the sample, chromatography is improved, and the risk of signal enhancement or suppression is generally reduced. Although it requires specialist equipment, the process can be automated offline or conducted entirely online prior to the HPLC stage of analysis.

Solid phase extraction cartridges are predominately differentiated by (a) sorbent material, (b) mass of sorbent material in the cartridge, and (c) cartridge size. The mass of sorbent material and the size of the cartridge selected largely depend on the volume of liquid to be analyzed. Larger-sized cartridges (up to 60 mL) are able to hold more water at any given time, reducing the demand on the analyst. Similarly, a larger mass of packing material (generally up to 1 g) provides more surface area for target analytes to bind, increasing the volume of the sample that can be loaded onto the cartridge-packing material. The size and sorbent material of the selected product should be optimized for respective analytical demands. Generally, 60 mg of adsorbent material is appropriate for 75–250 mL of water, while 200 mg can load up to 1 L (Waters Corp., 1998).

Solid phase extraction using any cartridge type follows the same general protocol (Fig. 2). An aqueous sample is first pretreated (if necessary) via filtration to remove coarse particulate materials or by pH manipulation. The adsorbent material is then conditioned with up to one cartridge volume of a water-miscible solvent (e.g., methanol) to wet the bonded phases. This ensures a consistent interaction between the analyte and adsorbent functional groups. Following conditioning, the same volume of reagent water is used to displace the solvent and condition the cartridge for the loading of the aqueous sample. The sample is then loaded into the cartridge using a vacuum at a rate of up to 5 mL/min. Slower sample loading ensures that the target analytes have more time to interact with the adsorbent material, thereby increasing chemical retention. After the sample is loaded, the adsorbent material is washed with up to one cartridge volume of water in order to remove any retained interfering matrix material. The cartridge contents are then dried under vacuum for approximately 20 min to remove excess water prior to elution. Elution of the target analytes is achieved by interrupting the interactions between the adsorption material and the target analytes. This can be done using 1–2 elution steps with an organic solvent (e.g., acetonitrile and/or methanol) or by the manipulation of chemical or adsorbent material ionization to

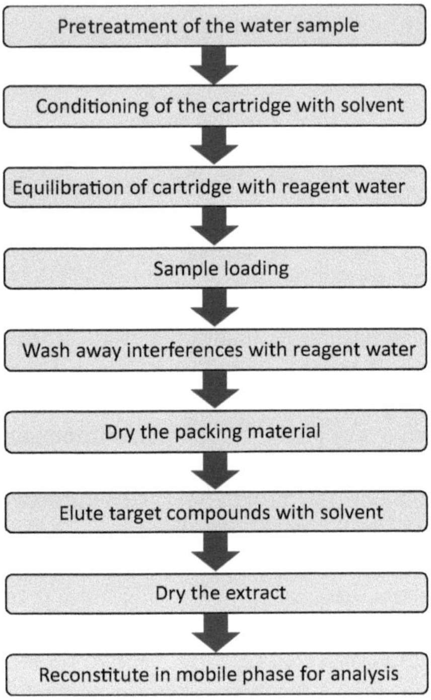

Fig. 2 General protocol used for the SPE of target analytes in an aqueous matrix.

weaken the analyte/adsorbent bonding. The final extract is then dried and reconstituted into a solution suitable for HPLC separation (e.g., starting mobile phase).

The sorbent material used in modern SPE applications varies from being highly selective to general purpose, largely selected based upon the physiochemical properties of the targeted analytes. For cleanup and enrichment of most organic contaminants in water, reversed-phase and ion-exchange SPE sorbent materials are employed (Fig. 3). Ion exchange is based on the interaction of the ionized target analytes and sorbent materials with functional groups of opposing polarity. These interactions are largely mediated by electrostatic interactions, which may be disrupted by various mechanisms including affinity to organic solvents and neutralization of the ionization in either of the target analyte(s) or sorbent material during elution. Silica particles amended with hydrocarbon chains typically between C8 and C18 in length (reversed-phase materials) offer the capability to retain analytes of moderately polar to nonpolar nature. This is predominately achieved via hydrophobic interactions and van der Waals dispersion forces, which are disrupted during elution via the chemical affinity for the organic solvent. Similarly, the so-called copolymer adsorbent materials contain both hydrophobic and hydrophilic monomers (e.g., divinylbenzene and vinyl pyrrolidinone) capable of retaining a wide range of polar and nonpolar analytes. Copolymer SPE is among the most commonly used materials for cleanup and enrichment of the organic contaminants in water due to their widely applicable nature. Hydrophobic/lipophilic materials are

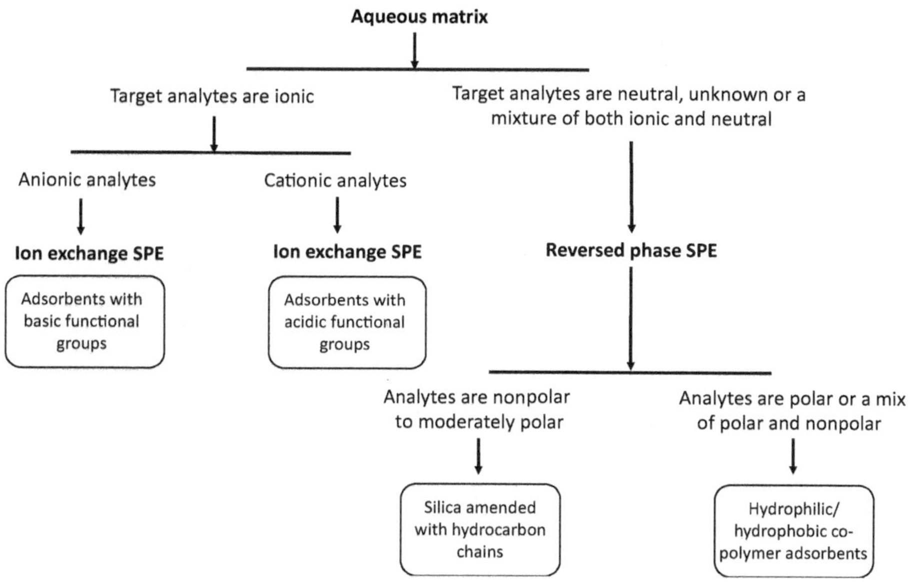

Fig. 3 Flow chart detailing the type of SPE material most commonly used for the analysis of organic contaminants in an aqueous matrix and applicable physiochemical traits of their target analytes.

generally useful for the selection of agrochemicals, pharmaceuticals, metabolites, and other organic contaminants in water (Waters Corp., 1998).

It is important to note, however, that target chemical ionization can affect the retention of analytes on reversed phase SPE cartridges. To alleviate the loss of ionized analytes during cartridge loading, manipulation of the sample pH (2 pH units above or below the pKa value of the target chemical(s)) may be used to ensure that target analytes are in their neutral form when loading the sample. This will increase hydrophobicity and retention to reversed-phase SPE media.

4 High-performance liquid chromatography

Prior to entering the vacuum of the mass spectrometer, target analytes in the sample must be separated from the mixture. This is efficiently accomplished using HPLC. The technique has been widely coupled to MS since the late 1900s for use in the analysis of organic contaminants in liquid matrices (Lindon et al., 2000). HPLC is based on the principle that chemicals within the sample injected into a chromatography column will interact with both the absorbent packaging material (typically silica particles or a derivation thereof) and the liquid flowing through the column itself based on their physiochemical traits at maximum pressures up to 500–600 bar.

Chromatography columns used for HPLC are differentiated by (a) length, (b) packing material/particle size, and (c) internal diameter. The most widely used columns for the analysis of organic contaminants in water are the so-called reversed-phase columns, comprised of silica particles with hydrocarbon chains of C8–C18 in length attached to the particle surface. Packaging with particles of smaller diameter will increase the surface area available for binding interactions during the run. Two liquid mobile phases are required: (A) a polar solvent such as water and (B) an organic solvent such as methanol or acetonitrile. The relative proportion of the mobile phase A to B flowing through the column, flow rate (typically up to 0.5 mL/min), and internal diameter of the column determine pressure. The mobile phase flow can be isocratic or operated as a gradient. While isocratic flow maintains a constant proportion of the two mobile phases, most HPLC for organic contaminants operates as a gradient of increasing organic phase proportion (e.g., acetonitrile or methanol) over the run duration.

An autosampler component of the HPLC instrument will aspirate an aliquot of the sample, which is then injected into the chromatography column. A typical injection volume ranges between 1 and 100 μL and is largely based on the method of sample pretreatment or enrichment. Once in the column, polar analytes will have a higher affinity to the mobile phase at the initial stage of the run than to the packaging material. Hence, polar molecules will be the first to exit the chromatography column in the aqueous-dominated mobile phase. During this time, nonpolar molecules will have partitioned to the hydrocarbon groups attached to the column-packing material via hydrophobic interactions and van der Waals dispersion forces. As the run progresses, the proportion of the organic mobile phase increases during gradient chromatography. Nonpolar analytes will leave the column once their interaction with the packaging material has been overcome by their affinity for the increasing dominance of the organic phase. The time at which an analyte leaves the chromatography column and is first detected by the mass analyzer is known as the retention time. Retention times are the first identifier of analyte identity due their specific mobile phase/adsorbent affinity affecting a reproducible elution from the column. After completion of the gradient, the mobile phase returns the original, aqueous-dominated proportion to allow the internal column pressure to re-equilibrate between runs (typically for 5–10 min).

As soon as an analyte leaves the column, it is directed via the mobile phase to the atmospheric pressure chamber of the mass spectrometer. Here, the mobile phase is removed, and the target analytes enter the spectrometer as an ionized gas. Constituents such as formic acid, ammonium formate, or citric acid may be added to the mobile phase flow to assist in the formation of ions entering the spectrometer. For example, the addition of formic acid (e.g., 0.01 M) assists in the ionization of neutral molecules by providing a source of protons, which form adducts with the analytes in the electrospray plume before entering the spectrometer vacuum (Furlong et al., 2014).

5 Mass spectrometry

The measurement of mass using a mass spectrometer is based on the principle that the flow of charged molecules will bend along an electrical or magnetic path and, hence, separate from one another. Multiple types of mass spectrometers are available for use with liquid chromatography, including time-of-flight (TOF), triple quadrupole, and ion trap spectrometers (Choi et al., 2003). This section will focus on triple quadrupole MS as this technique is commonly used for targeted quantification of organic contaminants in water after separation by HPLC. A brief discussion of TOF and ion trap MS is provided.

5.1 Source ionization and ion transfer

Target analytes in the liquid mobile phase leaving the HPLC must first enter the vacuum of the MS in the gaseous and ionized form in order to ultimately reach the detector. Furthermore, as water expands in volume upon evaporation, the mobile phase itself must be vaporized in a way that does not decrease the vacuum inside the MS. Such a vacuum interference may affect movement in the quadrupoles. Several ionization techniques are employed including electrospray ionization (ESI), chemical ionization, matrix-assisted laser desorption ionization, atmospheric pressure chemical ionization (APCI), and electron ionization. Atmospheric pressure ionization (API) techniques (e.g., ESI and APCI) allow both the removal of mobile phase and ionization of target chemicals outside of the MS vacuum and are predominately used for triple quadrupole MS. ESI is commonly used for the analysis of organic contaminants in water (e.g., pharmaceuticals) and is applicable to thermally sensitive chemicals across medium to high polarity and at a wide range of molecular mass. ESI occurs by the formation of a fine mist of charged droplets formed by the application of high voltage (3–5 kV) and a nebulizer gas at the tip of a fine capillary containing the mobile phase flow. As the mobile phase droplets evaporate, the droplet size decreases, thereby increasing the electric field on the droplet surface to the point where mutually repulsive forces exceed droplet surface tension and fission occurs. As this process continues, droplets eventually become small enough to allow the target ions into a gaseous phase. Chemical ionization in the resulting gas is by protonation (positive ESI) and/or deprotonation (negative ESI) reactions including proton transfer and electrophilic addition reactions. The ions are of the same polarity as the voltage applied at the nebulizer and both positive and negative ESI may occur simultaneously in most modern instruments.

Similarly, APCI also removes the liquid mobile phase and delivers target chemicals ionized in a gaseous phase to the MS. However, APCI evaporates the mobile phase via pushing the flow through a heater using a gas such as nitrogen, making the technique more applicable to thermally stable chemicals of low to medium polarity.

Once the target analytes are ionized, an ion capillary then directs their flow from the atmospheric pressure ion source compartment into an initial ion funnel. This funnel is operated at high pressure and directs forward movement (Z-direction) of the analytes into the spectrometer via a second, lower-pressure funnel. Ion optics (e.g., an octapole or radio frequency lens) then focus the beam of ions before entering the quadrupoles.

5.2 The quadrupoles

Quadrupoles are formed by four parallel cylindrical metallic rods carrying electrical currents in a vacuum. Two diagonally opposite rods exert a direct current, whilst the other rod pair exerts an opposite high-frequency alternating current (e.g., radio frequency). This causes ions to oscillate in the X and Y directions between the quadrupoles whilst still moving forward (Z-direction). By manipulating the alternating current, ions of a specific mass-to-charge (*m/z*) ratio can be oscillated to the point where they either collide with one of the quadrupoles or are ejected from the system. In so doing, the oscillation of targeted ions can be stabilized enabling their selective transfer into the next quadrupole. This quadrupole (Q) configuration is considered a "mass filter" in that ions with specific m/z ratios are specifically selected to continue toward the detector whilst other, nontargeted, species are removed.

In a triple quadrupole system, the first quadrupole (Q1) specifically filters the targeted precursor molecules (i.e., the protonated and/or deprotonated target analytes) through the quadrupole whilst removing any ions of other *m/z* ratios. Direct and alternating currents needed to stabilize the oscillation of target ions are optimized during method development in order to minimize the ultimate signal loss. Target ions are then transferred to the collision cell (Q2), which can be operated as a quadrupole or, in some instruments, a hexapole. Here, a neutral gas (e.g., argon) is accelerated into the precursor ions, which induces dissociation in an energy-dependent manner. The energy required to dissociate the precursor ion into ionized transition fragments is referred to as collision energy and is optimized for each analyte during method development. The third and final quadrupole (Q3) acts as a second mass filter. It receives the ion fragments produced in the collision cell (Q2) and is used to select fragments of desired *m/z* ratios. Typically for quantitative analysis, the two most abundant transition ions produced via the dissociation of respective precursor ions are selected in Q3. These two transition ions and the respective direct and alternating currents needed to stabilize them in the third quadrupole are also optimized during method development.

5.3 Electron multiplication and detection

After filtering for the desired transition ions in the third quadrupole, ion signals must be converted into usable data (De Hoffmann et al., 1996). From the final quadrupole in most modern mass analyzers, the transition ions are directed toward a high-voltage conversion dynode, which may then be followed by an electron multiplier. In MS, a dynode is an electrode operating within a vacuum that emits electrons as a result of an ion of sufficient *m/z* ratio and kinetic energy colliding with its surface. Here, the signal produced by this collision is amplified via secondary particle emission of electrons.

Both positively and negatively charged ions hit the high-voltage conversion dynode, which then emits an initial group of electrons. This is achieved via ionic attraction between the large applied voltage (e.g., positive or negative 15,000 V) of the conversion dynode and the transition ion charge. From here, the electrons may be further amplified via collision with a series of lower-voltage dynodes, each with resisters between them (electron multiplier). At each subsequent dynode collision, the electrons are multiplied exponentially while the voltage applied at each stage decreases until the last, grounded dynode is reached. The electrons present after the final grounded dynode represent the detected signal corresponding to the respective transition ions.

5.4 Quantitation and the calibration curve

Quantification of target analytes occurs via a linear response relationship derived by the MS signal for a known standard at various points of increasing concentration (the calibration curve) and the MS response for an analyte of unknown concentration. The identity of target analytes is determined by (a) LC retention time, (b) precursor ion mass, and (c) the ratio between abundance of the two transition ions. Of the two transition species, the ion with greatest abundance is used for quantitation. This can occur via either an external or internal calibration. External calibration is the simplest approach and is accomplished without internal standards using a calibration curve of standards for each target analyte. Accuracy can be ensured using external quality controls. An internal calibration relies on the ratio between the peak area of the target analyte relative to that of an internal standard at a constant concentration throughout the calibration curve. Here, each concentration level of the calibration curve has the same amount of an internal standard, which is structurally similar to the target analyte (if not the analyte itself) and is typically labeled with deuterium, ^{13}C, or ^{13}N. Rather than simply using peak abundance/height in an external calibration, internal methods use the ratio between the analyte abundance and that of the known internal standard of known concentration. Internal standard calibrations are more reliable and accurate than an external calibration; however, the cost of labeled standards may be prohibitive in some instances. Furthermore, the ability of a labeled chemical to act as a reliable internal standard for an analyte should be confirmed during method development. With either external or internal calibrations, the expected concentration of the unknown analytes should fall within the linear calibration range, which is typically formed by 6–15 respective concentrations.

5.5 Other mass analyzers coupled with HPLC: Time-of-flight and ion trap MS

Although triple quadrupole mass spectrometers are typically favored for the analysis of organic contaminants in water, largely due to their selectivity, other spectrometers are coupled to HPLC separation. These include TOF and ion trap mass spectrometers.

Time-of-flight mass spectrometers measure the mass-dependent time an ion takes to travel a set distance in response to an applied voltage. Key to this technique is the timed release of ions from the ionization source. Ionization can occur via pulses of voltage releasing ions in bursts. Matrix-assisted laser desorption ionization is another common ionization technique and is based on a laser (e.g., ultraviolent nitrogen laser light, $\lambda = 337$ nm) heating and vaporizing the outer surface of a material, which then releases ionized molecules into the spectrometer (Shimadzu, 2019). From the ionization source, ions are attracted in a vacuum to the grounded detector via a potential difference. Highly charged ions and those with smaller *m/z* ratios move faster along the vacuum chamber. The *m/z* ratios of the ions are then determined based on the time (i.e., speed over a set distance) they take for travel to the detector. TOF-MS instruments allow for the analysis of a wide mass range of analytes with near-simultaneous detection (Shimadzu, 2019).

Ion trap mass spectrometers are also often coupled to HPLC. These instruments are able to store ions in a three-dimensional quadrupole chamber using direct current and radio frequency fields after ionization. Operating with a similar ion selection mechanism as a triple quadrupole mass filter, by manipulating the electric fields, the spectrometer is able to select ions of a particular *m/z* ratio whilst ejecting others into the detector. Ions can be stored in the trap for milliseconds to days, although degradation and adduct formation may result in atypical mass spectra over prolonged periods. Using an inert gas, the chamber may also be used for multiple-stage ion fragmentation (i.e., MS^n), providing analyte structural data. Ion trap mass spectrometers allow for high mass resolution and sensitivity with applications in nontargeted and suspect-targeted screening. As the technology, and concomitant mass precision, of TOF and ion trap MS improves, their use in nontargeted and suspect-targeted analysis will grow (e.g., Kunzelmann et al., 2018; Martínez-Piernas et al., 2018).

6 Quality assurance and control measures

Due to the specific and sensitive capability of HPLC-MS/MS methodologies, quality assurance (QA) and quality control (QC) are of utmost significance to ensure the integrity of resulting data. Key QA/QC measures should be employed at every stage, from sample collection to data analysis. These include (but are not limited to) the collection of field blanks during sampling campaigns, use of internal standards in quantitative analysis, continuing calibration blanks and spikes (QCs), and robust method validation.

6.1 Field blanks

For the use of field blanks, the sample collection protocol is followed in the field with an aliquot of pure, reagent-grade water (e.g., LCMS grade), which does not contain residues of the target analytes. The field blank is then processed using the same procedure as environmental samples. This allows the analyst to determine the extent of interference or contamination of target analytes originating from the materials or collection protocol itself. Adjustments can be made to limit any method-derived interference (e.g., precleaning of sample vials with solvent, and use of gloves or a mask during sample collection).

6.2 Internal standards

The use of internal standards is marked by the addition of structurally similar or isotopically labeled chemicals of known and constant concentration. Use of internal standards in the calibration samples and/or spiked into the test samples can increase the accuracy and precision of the resulting data. An appropriate internal standard will behave in the same way as the associated target analyte in response to any interference. Therefore, the ratio between MS response to the target analyte and that of its internal standard will exhibit less variability than the response for the target analyte alone. Use of internal standards can also help correct analyte loss during pretreatment or storage.

6.3 Blanks within analytical sequences

A so-called blank used within the respective analytical sequences is the injection of pure (i.e., reagent-grade) water without residues of the target analytes. By injecting blanks within an analytical sequence (e.g., after every 10 injections), the analyst can determine (a) the presence or extent of crossover between injections, (b) contamination of the instrument, and (c) background interference.

6.4 Spiked standards within analytical sequences

These samples, sometimes referred to as continuing calibration verifications, contain known amounts of the target analytes in mixture. By injecting these solutions within an analytical sequence (e.g., after every 10 injections), the analyst can determine (a) the presence and extent of analytical drift, (b) accurate quantification throughout the analytical run (of particular importance during lengthy runs of multiple days), and (c) changes in sensitivity or background interference throughout an analytical run (e.g., indicating dirty optics or nebulizer occlusion).

6.5 Method validation

A method validation exercise is a robust study of method accuracy, precision, and repeatability. By conducting a thorough and robust validation of the analytical method (e.g., Furlong et al., 2014), the reliability of any data produced is ensured. Full-method validations can be extensive and span multiple days in more than one laboratory. However, the quality of the method and resulting confidence in data or research conclusions produced are more highly regarded following validation exercises.

7 Conclusions

HPLC-MS/MS has emerged as a crucial tool for the study of organic contaminants in the aquatic environment. Effective separation and specific quantitative capacities have enabled sensitive analysis of potentially ecotoxic chemicals at biologically relevant concentrations. The dominance of HPLC-MS/MS as a quantitative tool is likely to remain for a considerable time. With recent advances in nontargeted or suspect-targeted MS, advances in mass-accuracy and applicability of the technique are likely to drive new and wider applications of the technology in environmental science in the years to come.

References

Annesley, T.M., 2003. Ion suppression in mass spectrometry. Clin. Chem. 49 (7), 1041–1044.

Choi, B.K., Hercules, D.M., Zhang, T., Gusev, A.I., 2003. Comparison of quadrupole, time-of-flight, and Fourier transform mass analyzers for LC-MS applications. Curr. Trends Mass Spectrom. 18 (5S), 524–531.

De Hoffmann, E., Charette, J., Stroobant, V., 1996. Mass Spectrometry. Principles and Applications. John Wiley and Sons, pp. 94–97.

Furlong, E.T., Noriega, M.C., Kanagy, C.J., Kanagy, L.K., Coffey, L.J., Burkhardt, M.R., 2014. Determination of human-use pharmaceuticals in filtered water by direct aqueous injection—High-performance liquid chromatography/tandem mass spectrometry. 5 US Geological Survey Techniques and Methods, p. 49.

Gosetti, F., Mazzucco, E., Zampieri, D., Gennaro, M.C., 2010. Signal suppression/enhancement in high-performance liquid chromatography tandem mass spectrometry. J. Chromatogr. A 1217 (25), 3929–3937.

Kim, J.W., Jang, H.S., Kim, J.G., Ishibashi, H., Hirano, M., Nasu, K., 2009. Occurrence of pharmaceutical and personal care products (PPCPs) in surface water from Mankyung River, South Korea. J. Health Sci. 55, 249–258.

Kolpin, D.W., Furlong, E.T., Meyer, M.T., Thurman, E.M., Zaugg, S.D., Barber, L.B., Buxton, H.T., 2002. Pharmaceuticals, hormones, and other organic wastewater contaminants in US streams, 1999-2000: a national reconnaissance. Environ. Sci. Technol. 36 (6), 1202–1211.

Kunzelmann, M., Winter, M., Åberg, M., Hellenäs, K.E., Rosén, J., 2018. Non-targeted analysis of unexpected food contaminants using LC-HRMS. Anal. Bioanal. Chem., 1–10.

Lindon, J.C., Nicholson, J.K., Wilson, I.D., 2000. Directly coupled HPLC–NMR and HPLC–NMR–MS in pharmaceutical research and development. J. Chromatogr. B Biomed. Sci. Appl. 748 (1), 233–258.

Martínez-Piernas, A.B., Plaza-Bolaños, P., García-Gómez, E., Fernández-Ibáñez, P., Agüera, A., 2018. Determination of organic microcontaminants in agricultural soils irrigated with reclaimed wastewater: target and suspect approaches. Anal. Chim. Acta 1030, 115–124.

Neilson, A.H., 2017. Organic Chemicals in the Aquatic Environment: Distribution, Persistence, and Toxicity. CRC Press.

Roll, I.B., Halden, R.U., 2016. Critical review of factors governing data quality of integrative samplers employed in environmental water monitoring. Water Res. 94, 200–207.

Shimadzu, 2019. Principles of MALDI-TOF Mass Spectrometry. [online], available at https://www.shimadzu.com/an/lifescience/maldi/princpl1.html. last accessed: 15 April 2019.

Vrana, B., Allan, I.J., Greenwood, R., Mills, G.A., Dominiak, E., Svensson, K., Knutsson, J., Morrison, G., 2005. Passive sampling techniques for monitoring pollutants in water. TrAC Trends Anal. Chem. 24 (10), 845–868.

Waters Corp, 1998. Solid-Phase Extraction with Oasis® HLB Sorbent: Simple Procedures for Superior Sample Preparation. Waters Web Assets Library. [Pdf], available at http://www.waters.com/webassets/cms/library/docs/my_oas98.pdf. last accessed: 12 April 2019.

Wilkinson, J.L., Hooda, P.S., Barker, J., Barton, S., Swinden, J., 2016. Ecotoxic pharmaceuticals, personal care products, and other emerging contaminants: a review of environmental, receptor-mediated, developmental, and epigenetic toxicity with discussion of proposed toxicity to humans. Crit. Rev. Environ. Sci. Technol. 46 (4), 336–381.
Wilkinson, J.L., Boxall, A., Kolpin, D.W., 2019. A novel method to characterise levels of pharmaceutical pollution in large-scale aquatic monitoring campaigns. Appl. Sci. 9 (7), 1368.

Emerging technologies for autonomous in-situ monitoring of water quality

2

Andrew Donohoe, Dermot Diamond, and Margaret McCaul
Insight Centre for Data Analytics, National Centre for Sensor Research, School of Chemical Sciences, Dublin City University, Dublin, Ireland

1 Introduction

Reliable high-frequency and widely distributed data relating to the water quality of lakes, rivers, and oceans are essential for effective water management and for implementing strategies designed to improve the overall quality of water resources (Ballance and Bartram, 1998). Monitoring the concentration and composition of potential contaminants in natural waters over time in multiple locations is also necessary to characterize and understand the aquatic environment (Sequeira, 2002; Hou et al., 2013). Imbalances of nutrients or other potential contaminants such as heavy metals can cause major damage to aquatic ecosystems and water bodies (Jaishankar et al., 2014; Chapra et al., 2017).

Current approaches to water quality monitoring typically involve grab sampling at a restricted number of target locations and times. These samples are then stored and transported to a centralized analytical laboratory where they are subsequently analyzed using traditional methods such as colorimetry, ion chromatography, atomic absorption spectroscopy, or mass spectrometry (Williams et al., 2003). This analytical approach has a number of drawbacks. For example, while centralized instrumentation may be highly automated, sampling is highly labor-intensive, and there is considerable potential for sample degradation and contamination during collection, transport, or storage prior to analysis. Furthermore, this approach is time-consuming, leading to significant delays between sampling and reporting, and costly (i.e., it is non-scalable in terms of the number of locations sampled and frequency of sampling) (Prien, 2007; Duan et al., 2013a,b). This results in data with poor temporal and spatial resolution, providing at best only a superficial picture of the status of a water body. Sensors are devices capable of quantifying physical properties, chemical potential, or biological pathogens within a water body by means of an electrical response (Mehrotra, 2016). In comparison to the abovementioned traditional grab sampling methods, autonomous sensing platforms can generate higher-frequency in-situ measurements, with minimal risk of degradation of the target analytes or contamination of the sample, as the sample is measured directly on-site in real time.

Monitoring Environmental Contaminants. https://doi.org/10.1016/B978-0-444-64335-3.00001-3

The higher frequency of measurement coupled with reduced costs means that more devices can be deployed in situ, improving the temporal and spatial resolution of the data generated, which is especially important in areas where access is difficult for grab sampling (Herfort et al., 2016; Diamond et al., 2012). Scaled deployments provide a much more detailed picture of waterbody status dynamics, enabling rapid identification of a contamination event and source, and more effective implementation of control/remediation strategies (Sequeira, 2002). For in-situ autonomous sensing, reliability, and cost are the key factors determining scalability (Albaladejo et al., 2012). Recent improvements in autonomous sensor capabilities have the potential to provide a scalable model for high-frequency sampling strategies on a much larger scale. A number of autonomous water quality sensing platforms are currently commercially available, while others are still in early research and developmental stages. The most widely available and robust sensors for water quality monitoring are physical sensors followed by chemical and finally biological sensors (Goldman et al., 2007). As discussed later, this reflects the increasing challenges associated with the creation of autonomous sensing platforms along with the trend from physical to chemical to biological sensing (as discussed later).

Currently, a number of parameters such as pH, conductivity, and dissolved oxygen can be tracked using well-established sensors that provide real-time in-situ data (Kruse, 2018). However, monitoring of chemical contaminants such as nutrients requires more complex analytical methods and is still overwhelmingly based on manual sampling followed by laboratory analysis. Lab-on-a-chip technology has the potential to dramatically improve instrumentation for such environmental measurements (Jang et al., 2011). These are typically developed and tested in research laboratories but are rarely rigorously assessed within the real environment, due to the barriers that inhibit effective transfer of emerging technologies toward higher technology readiness levels. Improved robustness and reliability are essential for these methods, so that they can withstand harsh conditions that can often occur during real-world extended deployments (Jokerst et al., 2012). If these improvements can be delivered, cost-effective widely distributed sensing networks present enormous opportunities for the collection of rich data sets that, together with data from other sources such as satellite remote sensing, will provide us with a much clearer picture of the chemical processes that occur within water bodies (Goldman et al., 2007; Clements et al., 2017; Jones et al., 2019).

2 Water quality monitoring

To correctly manage water resources, it is essential to monitor water quality status with appropriate temporal and spatial resolutions for accurate modeling. Typically, this requires a multiparametric approach encompassing pH, temperature, conductivity, dissolved oxygen, alkalinity, nitrogen, phosphorus, organic compounds (organic carbon, humic substances), microorganism profiling, and biologically active compounds (algae, phytoplanktons). To cover all of these is beyond the scope of this chapter, and we will focus on nutrient monitoring as an exemplar, as this will encompass issues and potential solutions that apply to many other analytes.

A number of EU and global policies and directives are driving increased demand for autonomous sensing platforms, as current approaches cannot provide the means to scale up deployments toward the required temporal and spatial resolution. The EU Water Framework Directive (WFD) aims to prevent further deterioration in the status of surface waters, groundwater, and water-dependent ecosystems and aims to restore noncompliant water bodies to at least "good status" (Gordon, 2016). The five water categories under the water framework directive are river, lakes, groundwater, transitional, and coastal (Environmental Protection Agency, n.d.). This directive is further supported by the Nitrates Directive, the Urban Wastewater Directive, the Drinking Water Directive, and the Groundwater Directive. The Water Framework monitoring programs are integral parts of the new management strategy for water to meet the "good status" aim. This can be achieved through the establishment of integrated monitoring programs that are capable of collecting physical, chemical, and biological data, in order to provide for the assessment of water quality status of both surface waters and groundwater. Through its International Initiative on Water Quality (IIWQ), the United Nations Educational, Scientific and Cultural Organization (UNESCO) has launched a comprehensive worldwide water quality portal based on remote satellite earth observational data which deal with freshwater systems, lakes, and rivers. The aim is to use the portal to assist in global water quality assessment, while also providing an educational resource for environmental managers, politicians and scientists on water quality from a global perspective (International Initiative on Water Quality: Promoting Scientific Research, Knowledge Sharing, 2015).

3 Sensing challenges: Platform and deployment hierarchies

In order to be successful, future autonomous environmental sensing platforms must meet a number of critical requirements. For example, they must be relatively inexpensive to buy and maintain so that they provide a route toward larger-scale deployments. This in turn demands that they can function autonomously for longer periods between servicing intervals, ideally up to several months. Throughout the periods between servicing, they must be capable of providing reliable continuous analytical data via wireless communications to cloud databases. Despite the increasing range and diversity of techniques currently available, continuous, online measurement systems remain largely limited by factors such as access to power, short service lifetimes, the need for frequent calibration, interferences, and fouling problems. These challenges must be considered during the development and deployment of more effective autonomous platforms.

In recent years, a number of initiatives have been launched with the specific aim of driving down the cost of in-situ sensing. For example, the Alliance for Coastal Technologies "Nutrient Sensor Challenge" (Bourgeois et al., 2003) provided significant awards and access to potential markets for technologies with a unit cost of $5000 or less, compared to the typical $20,000+ for such devices (Alliance for Coastal Technologies - Nutrient Sensor Challenge, n.d.; Pellerin et al., 2016). This nutrient challenge was initially introduced in 2014, with a second stage called in 2018 by the United States Environmental Protection Agency (EPA). The second stage required applicants to demonstrate how the data generated from in-situ sensing instruments could be used to inform decisions and actions, enhance nutrient management strategies, and inform communities of local water status. Such initiatives can drive competition in sensor development and help reduce sensor costs. The aim is that eventually, unit costs reduce to a "tipping point" that leads to larger uptake by state agencies and industry, thus increasing the size of the overall market.

The cost and complexity of a sensing platform (and as a result its scalability) depend both on the type of analyte and on the deployment scenario (see Fig. 1). Physical parameters (e.g., temperature, turbidity, conductivity, and color) are the least complex to track due to the availability of very reliable, low-cost sensors. As a result, these sensors can be deployed within the environment at a large scale. Physical sensors are typically low power, allowing long-term deployments without the need for servicing or maintenance.

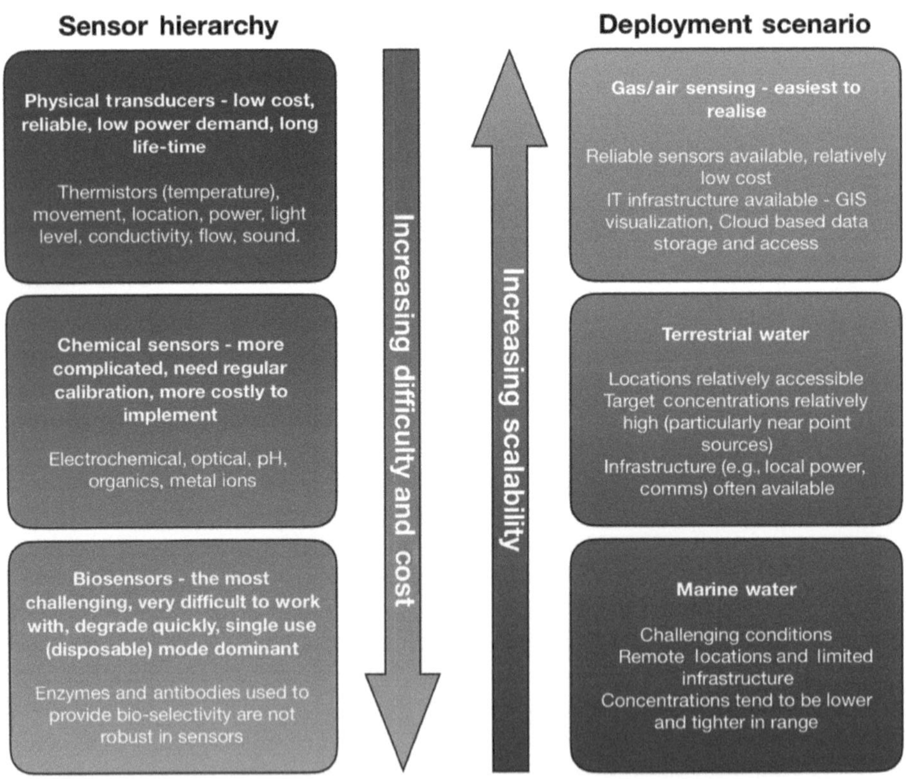

Fig. 1 Remote continuous sensing challenges, platform, and deployment hierarchies.

Chemical analytes (e.g., nutrients and heavy metals) often require complex detection methodologies and sophisticated sensing platforms to provide accurate in-situ data. The need for regular calibration is a major issue, as this requires a fluidic system to enable calibration and sampling lines to be switched. Calibration is required because of stability issues arising from changes in the surfaces of fluidic channels, optical detection windows, or functionalized sensors that are in contact with the sample. Introducing calibration also requires the storage of standards and retention of waste within the fluidic-based instrument. Furthermore, in addition to the requirement for reliable power and electronics common to all environmental sensing systems, reagents, and standards used in fluidic chemical analyzers must be stable for the duration of the servicing interval.

Biological analytes remain the most challenging sensing methods due lifetime and ruggedness issues associated with biosensors and biomaterials used in bioassays. Antibody-based measurements, while highly selective, are typically single use, due to the irreversible nature of the binding, making the provision of multiple time-series measurements extremely challenging. In general, enzymes degrade quickly and are incapable of working reliably over long periods of time in environmental deployments. Furthermore, complex environmental matrices can produce interferences that inhibit the detection of the target analyte (Nigam and Shukla, 2015).

In addition to the position of the sensing technology in the hierarchy, the deployment scenario must also be taken into account. Generally speaking, gas/air sensing is the most accessible deployment scenario, and the marine environment the most challenging, particularly far offshore ocean deployments (see ARGO, Fig. 2). Between these lies on-land freshwater monitoring.

The matrix presented in Fig. 1 is a useful guide for assessing the risk associated with a sensing deployment scenario. For example, deploying physical sensors to track air quality is the least risky combination, and this is reflected in the examples of such deployments worldwide (Bouabana et al., 2016; Singh and Singh, n.d.; Young et al., 2014; Mota et al., 2018; Lazarescu, 2015) that are already providing data with high spatial and temporal resolutions (Morawska et al., 2018). Increasing effort from scientists and instrument manufacturers as well as improvements in wireless technology has made it possible to reduce the cost of these sensors by up to three orders of magnitude, and as a result, broader utilization of information on air pollution and weather data has become possible (Kumar et al., 2015). For example, Popoola et al. (Popoola et al., 2018) presented the use of these low-cost air monitoring networks to quantify air quality in urban settings at London Heathrow Airport using low-cost sensing, but the approach could be applied to a wide range of environmental deployment scenarios.

In contrast, the deployment of biosensors in the marine environment is very high risk, with virtually no chance of success. Therefore, it is no surprise that marine deployments are focused overwhelmingly on monitoring physical parameters such as temperature, depth, flow, color, and turbidity, with chemical-sensing activity happening to a much lesser extent. This is exemplified in the ARGO project, which is a global network of almost 4000 profiling floats that provide continuous measurement of multiple parameters in the ocean (Argo, n.d.; The Argo Program Observing the Global Ocean with Profiling Floats, n.d.). These floats are sited at fixed depths

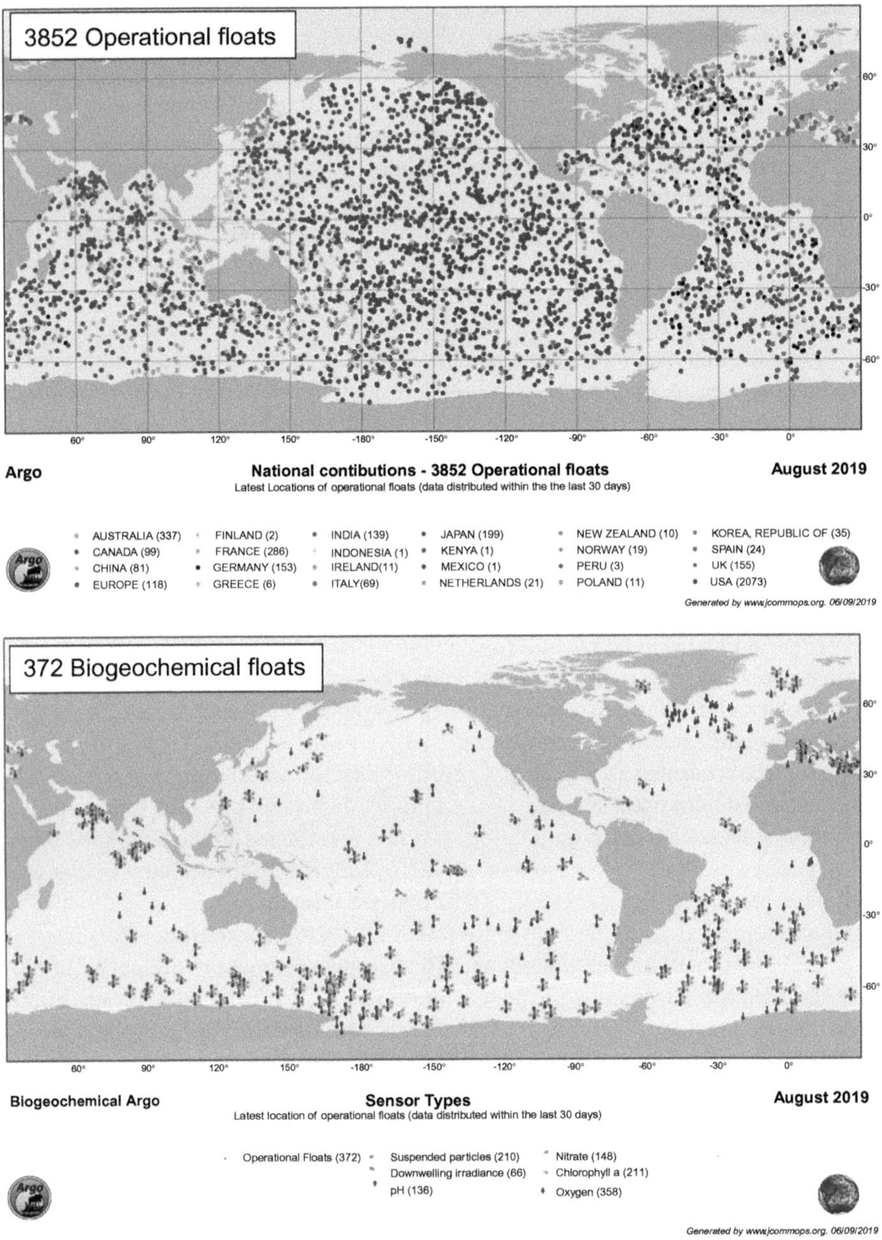

Fig. 2 Map of distributed total operational sensing floats (3852 top) under the ARGO project versus operational biogeochemical floats (372 bottom) (Argo, 2019) accessed September 2019, credit http://argo.jcommops.org.

(typically 1000–2000 m) and are designed to surface at 9-day intervals. While surfacing, they autonomously measure temperature and salinity. Upon researching the surface, they transmit their accumulated data via satellite.

Fig. 2 shows the number of currently deployed floats (3852) within the ARGO project. Of these, only 372 (9.7%) floats produce data on what are referred to as "biogeochemical parameters," with 148 (3.8%) of these providing nitrate data (by direct UV absorbance at 220 nm), 211 detecting chlorophyll (by color measurements), 136 providing pH data (3.5%, reagent-based measurements), and 358 measuring DO (9.3%, Clark-type cells) (Argo, 2019) (monthly data, accessed in September 2019). It should be appreciated that the nitrate and chlorophyll measurements, while they report important chemical and biological information, do not employ chemical-sensing/biosensing methods. This is clear evidence of the complexity and unreliability of chemical-sensing and biosensing methods inhibiting their use in challenging deployment scenarios. Furthermore, the marinized Clark-type cells used for the DO measurements have to withstand high pressures and very hostile external ocean environment, and consequently cost ca. $60 K each. Even with complex corrections for effects arising from variable temperature, pressure, and salinity on the signal, "calibration of the DO measurements by the SBE-IDO sensor remains an important issue for the future." (Thierry et al., 2018)

In Fig. 1, chemical-sensing and freshwater deployment scenarios fall between these extremes in terms of the complexity of the measurement technology and challenging deployment scenarios. It is understandable, therefore, that this combination is the focus of considerable research activity, as it provides an intermediate balance between risk and potential reward, and consequently this is where new technologies are most likely to emerge and be adopted. The emergence of a number of these new chemical sensors for freshwater and marine sensing has been highlighted by stage two of the Alliance for Coastal Technologies "Nutrient Sensor Challenge." The National Oceanography Centre (NOC), Southampton (Performance Verification Statement for the NOC Phosphate Analyzer, n.d.; Performance Verification Statement for the NOC Nitrate Analyzer, n.d.), Sea-Bird Scientific (Performance Verification Statement for Sea-Bird Scientific HydroCycle-Phosphate Analyzer, n.d.), and Systea (Performance Verification Statement for the Systea WIZ Probe Nitrate Analyzer, n.d.; Performance Verification Statement for the Systea WIZ Probe Phosphate Analyzer, n.d.) all partook in performance verification during stage two where a series of protocols were assessed using both reference samples and environmental deployments.

4 Stability challenges with chemical sensors

Most chemical sensors reported in the literature arise from research performed by academic groups, with little attention given to their performance under real-use scenarios. A sensor that has been tested only within a laboratory requires a significant amount of in-situ testing to determine its operational limits, how long it can function acceptably in the field, and the appropriate deployment strategy for the sensor. This extends to commercial chemical sensors designed for laboratory use (Ho et al., 2005). For example, ion-selective electrodes (ISEs) under laboratory conditions are capable of providing rapid analysis of pH and a large number of cationic and anionic species. The ISEs are generally considered less applicable platforms for in-situ environmental monitoring because they require frequent calibration due to signal drift, degradation of the electrode membrane, and reduction of their effectiveness due to biofouling of the active sensing surface.

Anastasova et al. (Anastasova et al., 2012) presented a solid-contact ISE for environmental monitoring of lead in water. The limit of detection in the nanomolar range made the electrodes favorable for the detection of low-level Pb^{2+} in environmental waters. Low-cost ISEs are applicable for single use or short use for on-the-spot testing of Pb^{2+}. However, continuous exposure to contaminated river water caused rapid changes in the sensor research characteristics, with the entire response function shifting by ca. +30 mV after one day (i.e., approximately an order of magnitude error; see Fig. 3). This type of instability means that these simple devices must be incorporated into a sophisticated fluid-handling system to enable regular calibration and fault checking, if they are to have any chance of success in a remote deployment.

In this paper, the possibility of including arrays of sensors was highlighted as a possible route to achieve long-term deployments, by using each for a short period of time, within which they hopefully remain within calibration, and then switching in successive electrodes sequentially. The suggested arrays would have the sensors incorporated into microfluidic platforms that would control the sensor conditioning, perform sample calibration or standard additions, and restrict the sensor period of exposure to real samples. Similar to biosensing (which also suffers from short-period or single-use and shot-use models), this method would have significant restrictions due to the required complexity of the fluidic system and associated cost. For instance, in order to use an array, the sensor would have to incorporate a series of valves into the sensing platform to allow sequential selection of each sensor. Therefore, in order to

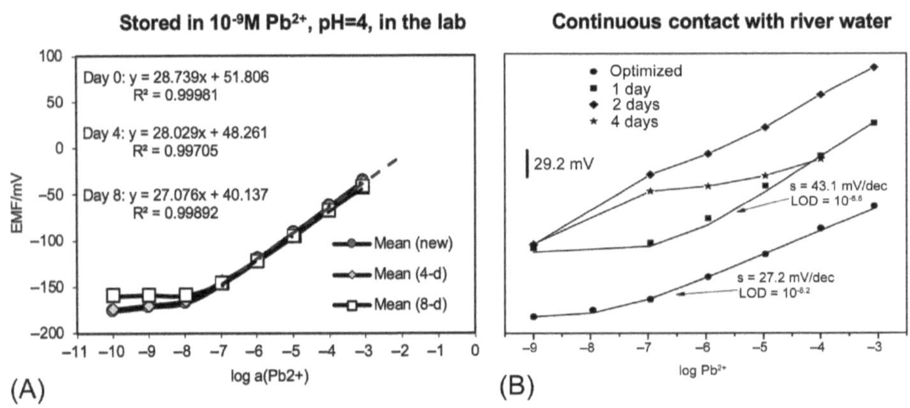

Fig. 3 Calibration data for the mean responses of disposable solid-contact ISEs for environmental monitoring of lead stored in (A) 10^{-9} M Pb^{2+}, pH = 4 in laboratory and (B) continuous contact with river water (Anastasova et al., 2012).

make this a feasible strategy, new approaches for creating fully integrated, low-cost, reliable valve structures within fluidic systems are needed (Delaney et al., 2017).

5 Influence of microfluidics

As mentioned earlier, microfluidic technology has tremendous potential to enable the integration of increasingly complex analytical procedures into autonomous instruments that are capable of mimicking laboratory-based methodologies with increased sensitivity and reduced analyte consumption. The concepts underlying microfluidics took inspiration from the tremendous success of integration and miniaturization in electronics in the 1960s and 1970s (Manz et al., 1990). Microfluidics typically utilizes microchannels with diameters in the range of 10 to 500 μm created in polymer, glass, or silicon substrates, to carry out reactions on a microscale by the manipulation of liquid samples and reagents (Czugala et al., 2011). Low dead volume is advantageous as it minimizes reagent consumption and waste generation. This minimization of fluid volumes used per assay dramatically reduces power requirements. As a result of the reduced reagent/sample volumes and power consumption, microfluidics provides a route toward the realization of platforms capable of longer deployment times in comparison to macroscale devices. The use of microfluidic technologies, therefore, is a key element in the development of compact low-cost sensing platforms, particularly when paired with low-cost detectors. Some examples are presented in Fig. 4.

However, there are also a number of challenges or disadvantages associated with the use of microfluidics. For example, due to the microscale channels, they are susceptible to blockages or interference caused by suspended solids during measurements. This can be mitigated by using wider channels (but this also increases dead volume) and implementing prefiltration of the sample prior to analysis (but this limits the target analytes detectable to only the dissolved fraction of the sample).

Measuring only using microlitres of the sample also has implications in terms of representative sampling, as small volumes of sample are often assumed to correlate with large water bodies, which may be highly heterogeneous in nature. However, this is more than compensated by the ability to take higher-frequency in-situ measurements at multiple locations (due to lower unit cost and increased scale), which will provide a much-improved description of the waterbody heterogeneity and dynamics.

6 Manual single-use sensing

Manual grab sampling still remains a standard methodology for water sensing where samples are obtained and brought back to the laboratory for subsequent analysis. Once sample collection, transport, storage, and analysis are carried out correctly, this method generates high-quality, highly reliable data. However, the cost per sample is high due to the highly manual nature of this approach and the time required to harvest and transport samples to the centralized facility. As a result, a number of on-the-spot tests for in-situ sensing have been developed to remove the need for storage and

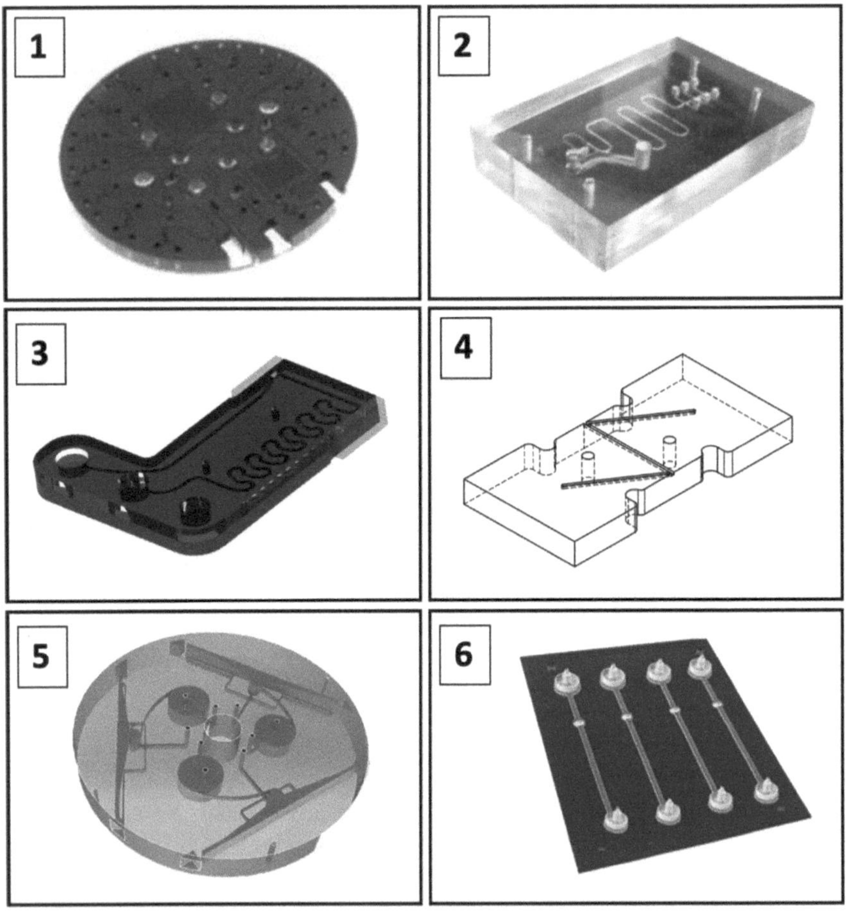

Fig. 4 Examples of microfluidic prototype technology used in sensing platforms. (1) PMMA layer of laboratory on chip for phosphate detection (Grand et al., 2017), (2) microfluidic chip for the direct determination of nitrate (Cogan et al., 2015), (3) opaque fluidic chip with optically clear windows for nutrient analysis (Donohoe et al., 2018) (4) z-shaped optical microfluidic detector, render adapted from Murray et al. (Murray et al., 2019), (5) PhosphaSense laboratory on disk for phosphate analysis (Duffy et al., 2017), and (6) flow control microfluidic chip-containing in-situ photoresponsive hydrogel valves (Delaney et al., 2017; Coleman et al., 2017).

transport back to the laboratory (Prien, 2007). These single-use tests are usually cost-effective per sample but still require manual user input and potentially require pre-treatment stages before results can be generated in the field.

In an attempt to increase the spatial and temporal resolution of single-use measurements, initiatives like the Earthwatch Institute Water Blitz (Hunston, n.d.) and Freshwater Watch (Earthwatch FreshWater Watch, n.d.) have been initiated. These initiatives encourage volunteers to carry out local sampling and analysis for key parameters such as phosphate and nitrite. The strategy of using trained volunteers provides a way to generate widely distributed data set that complements data acquired by more conventional means. Freshwater Watch is a global program developed by Earthwatch, which has gained over 100,000 volunteers worldwide, contributing to 10,000,000 h of data collection. Results can be logged on-site using a mobile phone app and can be visualized using its website (see Fig. 5). Similarly, the Earthwatch Institute carries out Water Blitz through which a drive for sampling is carried out on specific dates so that water quality can be directly correlated across Europe at the same time. The most recent Water Blitz aimed to sample in Dublin, Paris, and Luxembourg with volunteers sampling over a 3-day period (September 20–23, 2019). During this Blitz, volunteers sampled water for nutrients (nitrate and phosphate) using simple colorimetric kits and a visual color chart, turbidity, and color. A visual inspection of the sampling site is also logged to assess the surroundings, bank vegetation, and any potential sources of pollution in the immediate area.

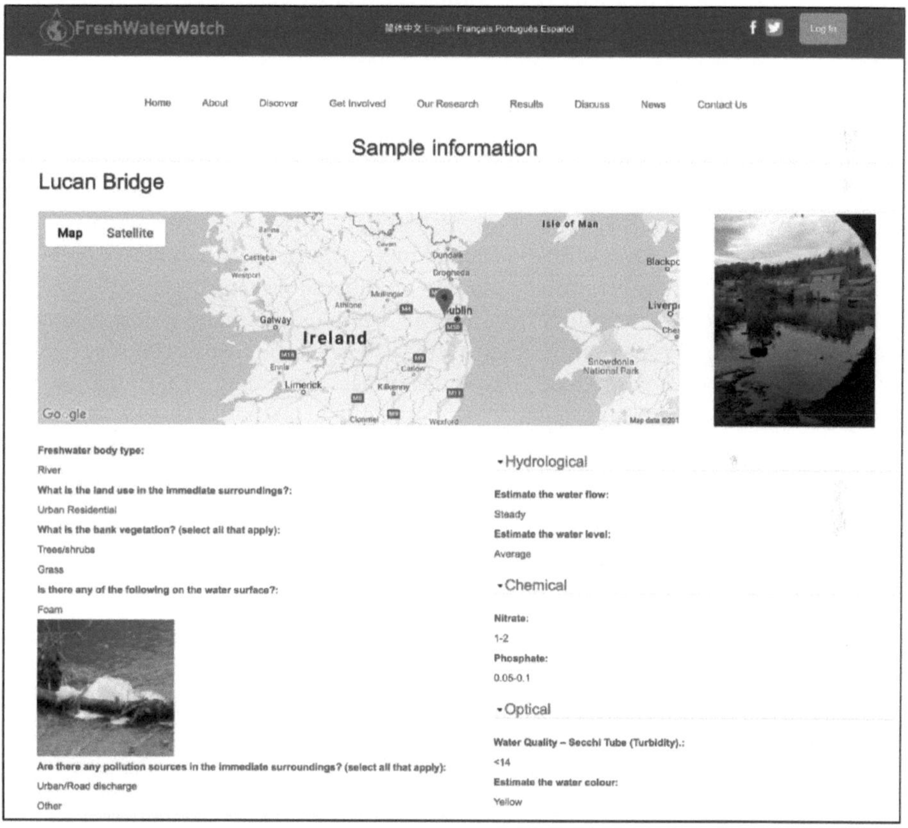

Fig. 5 Freshwater Watch interface showing sample data input by a volunteer citizen scientist including location, site image, and hydrological, chemical, and optical results (Lucan Bridge-23-08-2019-0, n.d.).
Credit Earthwatch, Citizen Scientist: Kevine – Friday, August 23, 2019.

Handheld colorimeters (DR900 Multiparameter Portable Colorimeter, n.d.) provide a quick and simple way to perform environmental water quality measurements in the field. Other methods include novel portable microfluidic technology such as that developed by Duffy et al. (Duffy et al., 2017) (see Fig. 4 image 5). This novel microfluidic disk is capable of in-situ quantitation of phosphate in natural waters using the molybdenum blue ascorbic acid colorimetric method.

7 Autonomous in-situ sensing for nutrients

Nutrients such as phosphorus and nitrogen are essential for plant growth and therefore for human life. However, the effect of population growth, agricultural practices, industrial processes, and pollution events has led to an excess of these nutrients natural waters. Excessive nutrient concentrations in water bodies can lead to eutrophication, which can cause release of algal toxins, formation of dead zones, and contamination of drinking water (Tchounwou et al., 2012; Yool et al., 2013; Heisler et al., 2008; Smith, 2003). Nutrients causing eutrophication can arise from anthropogenic and naturogenic effects (Yang et al., 2008) and can enter water bodies via point and diffuse sources (Bol et al., 2018). Point sources include direct releases from industry sites, wastewater treatment plant effluent pipes, and agricultural waste streams, whereas diffuse contamination of a water system occurs at multiple locations along its boundaries due to, for example, runoff from fields (Davidson et al., 2014). The diversity of these sources can make accurate profiling a challenging task. As a result, nutrient pollution still remains one of the most problematic environmental issues.

Eutrophication can drastically affect humans, animals, and plant life, and remediation is expensive, encompassing treatment costs for the removal of nitrogen, algal toxins and decomposition products. They also impact communities through reduced value for waterfront dwellings, reduced recreational and amenity value of water bodies, and economic losses for the tourism industry (Pretty et al., 2003). The economic cost of these events is estimated to be over $2.2 billion annually (Dodds et al., 2009) in the USA and £160 million annually in England and Wales (Pretty et al., 2003). Mitigation measures are difficult to implement without an accurately informed temporal and spatial model of the concentration profiles for nutrients within the water body (Nutrient Pollution - An Overview, n.d.). Therefore, higher-frequency nutrient measurements taken at a larger number of locations are essential to effectively combat this truly global problem.

One way to achieve this is by the use of autonomous in-situ sensing platforms to continually monitor the concentration of nutrients at varying locations within a water body. Autonomous sensing platforms are typically deployed in a fixed position during the sampling lifetime and aim to measure nutrient concentrations for extended periods ranging from months to years, with planned servicing intervals during their lifetime. This deployment methodology allows for high-frequency data at a single-point source at various locations. For long-term monitoring using autonomous in-situ sensing it is favorable for a platform to have the following characteristics:

1. Highly precise data with low detection limits
2. Fast measurement response time with good reproducibility
3. Stable long-term operation
4. Antifouling resistance

Sensor costs often limit the number of platforms available to deploy at a specific site. As a result, a compromise must be made for high-frequency data with good temporal resolution but with low spatial resolution per sensor deployed. The following section reviews the autonomous sensing platforms for the detection of target nutrients contributing to eutrophication that are either commercially available or currently under development.

7.1 Phosphorus

Phosphorus is an essential element for all life. It occurs in plants, microorganisms, and animal waste; it is also widely used in agricultural fertilizer and detergents (Khan and Ansari, 2005). Phosphorus is one of the limiting nutrients in eutrophication (Bowden et al., 2002). In water, phosphorus is present in a number of chemical forms with the most common being phosphates of which the inorganic orthophosphate form is of the most stable. Phosphates occur in a number of forms, orthophosphates, pyrophosphates, metaphosphates, polyphosphates, and organic and inorganic phosphates (Phosphate - An Overview, n.d.). Phosphate is mainly produced by anthropogenic processes and enters water bodies in a number of forms. Orthophosphates enter mainly through agriculture through runoff of soluble fertilizers from fields (Brogan et al., 2001). Organic phosphates enter water through biological processes associated with food and wastewater treatment. Condensed phosphates enter water through water treatment and through laundering and cleaning processes, although the latter is decreasing in importance due to the phasing out of phosphorus-containing detergents and cleaning products (Bowden et al., 2002).

7.2 Phosphate-sensing methods

Electrochemical methods are generally inexpensive, but phosphate electrochemical sensors have limited selectivity (due to chloride, bromide, iodide, and sulfide cross-sensitivity) and are prone to signal drift and long-term stability issues (Bresnahan et al., 2014; Fay et al., 2011). Phosphate can be detected electrochemically using potentiometry and indirect voltammetric detection based on the reaction of phosphate with a number of metals and associated complexes (Chen et al., 2018).

Fluorescence-based detection typically has higher sensitivity and lower limit of detection than absorbance-based colorimetric methodologies. Fluorescent detection of phosphate can be achieved by the replacement of a competing florescent ion within a complex with phosphate, which can then be detected optically. To date, this technique has been applied to flow through systems but while it has the potential to be integrated into an autonomous system in a similar way to colorimetric-based systems, this has not been reported to our knowledge. Frank et al. (Frank et al., 2006) proposed a method for the determination of dissolved reactive phosphorus using sequential injection analysis with the fluorescence quenching of rhodamine 6G-molybdate. This detection method had a limit of detection of 0.05 μM (Frank et al., 2006). This sequential injection system was used for the analysis of surface water in the North Sea, Wadden Sea, and Elbe Estuary. A reverse flow injection method using the same chemistry achieved a limit of detection of 7 nM for phosphate in seawater (Kröckel et al., 2014). Similarly, a flow injection method with chemiluminescence detection based on luminol oxidation by 12-MPA has also been used to determine phosphate in freshwater (Gimbert et al., 2007).

The most common method for the determination of phosphate in water is the use of colorimetric chemistries coupled with a light source (LED) and photodetector (photodiode) to measure the absorbance of a distinct color that results from a reaction between phosphate in the sample and a selective colorimetric reagent. The use of colorimetric chemistries has been optimized for use in flow-through systems and is favorable due to their selectivity, sensitivity, and reagent lifetime for some methods. The density of the color produced is related to the concentration of the target analyte using the Beer-Lambert law (($A = \log^{10}\left(\frac{I_0}{I}\right) = Lc$)) (Swinehart, 1962). When implementing these methods in autonomous instruments, reagent storage, stability, and waste containment must be considered, along with high method sensitivity to orthophosphate (large molar extinction coefficient, ε), broad linear calibration range, and reagent stability (Radu, 2009; Cleary et al., 2008). Fig. 6 shows a conceptual schematic for two common colorimetric reactions for phosphate detection, vanadomolybdophosphoric acid method, and the ascorbic acid method. The vanadomolybdophosphoric acid method (yellow method) for the detection of phosphorus is based on the colorimetric reaction of PO_4^{3-} with acidified vanadomolybdophosphoric that produces a yellow-colored solution detectable in the UV range below 400 nm. It is a simple colorimetric technique where under acidic conditions orthophosphate reacts with molybdate ions to form molybdophosphoric acid. In the presence of vanadate ions, this complex will form the strongly yellow-colored vanadomolybdophosphoric acid. This method is widely used due to the long-term stability of the reagent (at least 12 months) and its ease of integration to a microfluidic chip (Bowden and Diamond, 2003). The molybdenum blue/ascorbic acid method has been widely used for the determination of orthophosphate in environmental waters due to its high sensitivity and selectivity for orthophosphate. However, the method has a tendency to form precipitates in the final stage (To and Randall, 1977) that can cause issues in microfluidic systems such as clogging or changing flow characteristics within the chip itself as reported by Sequeira et al. (Sequeira, 2002). The reaction takes place in two stages, with the first stage involving the formation of a Keggin ion around the analyte anion. The second stage involves the reduction of this heteropoly acid to form a deeply blue-colored product. The resulting blue-colored complex can be detected spectroscopically between 660 and 880 nm. Although the vanadomolybdophosphoric acid method has a lower sensitivity with a low molar extinction coefficient ($\varepsilon = 8.38 \times 10^3$ mol l^{-1} cm^{-1} at λ_{max} 380nm (Bowden, 2003)) in comparison to the blue method ($\varepsilon = 2.61 \times 10^4$ mol l^{-1} cm^{-1} at λ_{max} 820 nm (Katewa and Katyare, 2003)), this is counterbalanced by its excellent

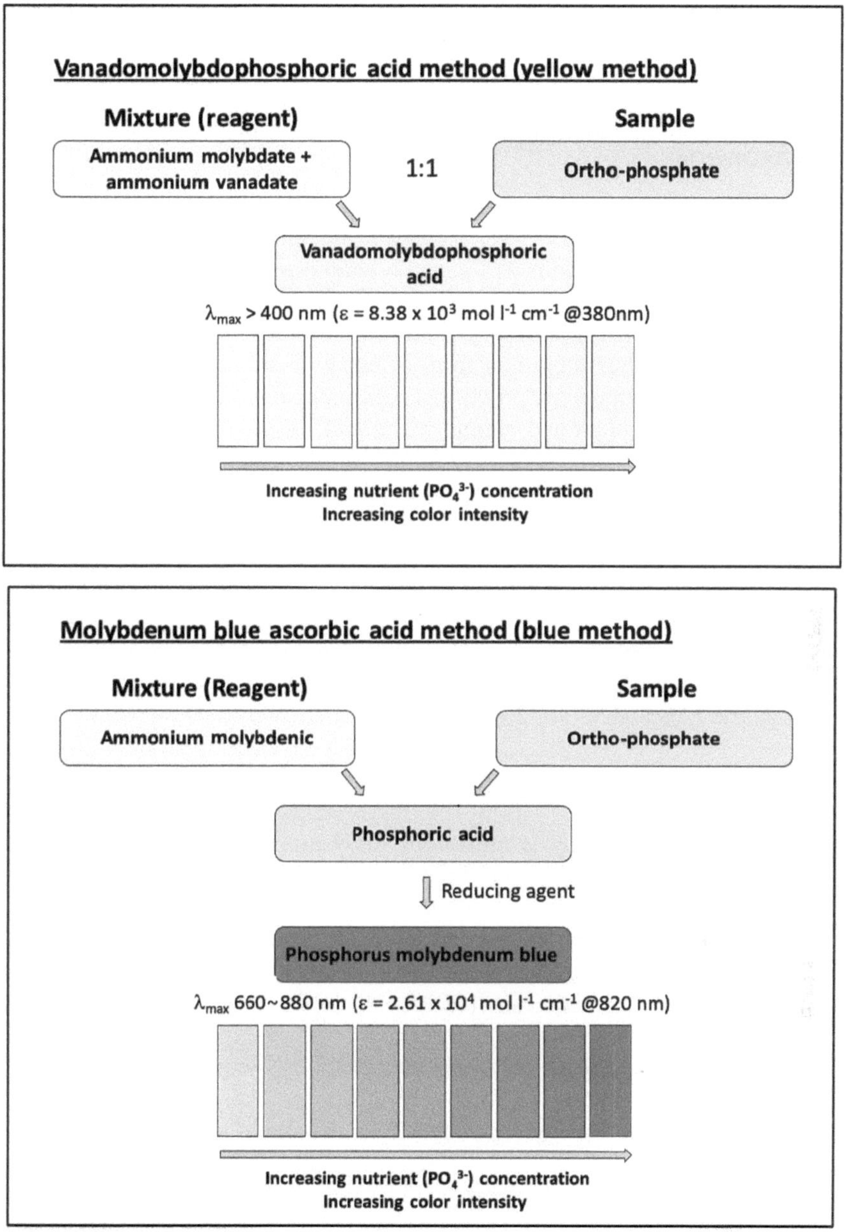

Fig. 6 Conceptual reaction schematic of standard colorimetric methods for phosphate detection: (Top) vanadomolybdophosphoric acid method (yellow) (molar extinction coefficient (Bowden, 2003)) and (Bottom) molybdenum blue ascorbic acid method (blue method) (molar extinction coefficient (Katewa and Katyare, 2003)).

reagent stability (up to 1 year) (Diamond et al., 2011). In contrast, the molybdenum blue ascorbic acid method is more suited for use on in-line sensors, as it has a high sensitivity but suffers from poor reagent stability (Sequeira, 2002; Duffy and Regan, 2017).

7.3 Nitrogen

Nitrogen is present in water in a number of chemical forms, the most common being nitrate (NO_3^-), nitrite (NO_2^-), organic nitrogen, and ammonia (NH_3) (Nordin, 2009). Nitrite and nitrate concentrations within natural and wastewaters vary widely. Nitrate is found in a number of natural processes such as during nitrogen fixation by cyano- and azobacteria and during soil degradation (Biological Nitrogen Fixation - An Overview, n.d.). The most common form of entry of nitrate into natural waters in excess is through anthropogenic processes. Anthropogenic sources of nitrate include agricultural fertilizers, plant and animal waste and wastewater discharges (Kaushal et al., 2011). Elevated levels of nitrates in natural water systems pose a risk to both the environment and human health (Ward et al., 2018).

Ammonia is a naturally occurring form of nitrogen in the nitrogen cycle. Dissolved ammonia gas is toxic to aquatic life in concentrations as low as 25 $\mu g/l^{-1}$[92]. Ammonia ranges in concentration from as low as 10 μg/l in surface and groundwaters and up to 30 mg/l in wastewaters (Cogan et al., 2014). Ammonia enters the aquatic environment in excess through anthropogenic sources such as agricultural runoff and wastewater discharged and is also released during nitrogen fixation and the breakdown of animal and plant waste (Absalan et al., 2004; Cogan et al., 2014).

7.4 Nitrite and nitrate colorimetry

The colorimetric Griess assay is most commonly used for the detection of nitrite. Fig. 7 shows a conceptual schematic for colorimetric reactions for nitrite detection. The reaction is based on the diazotization reaction of nitrite with sulfanilamide and N-1-naphthylethylenediamine dichloride (NEDD) under acidic conditions, which results in the formation of a very intense pink azo dye that can be detected spectroscopically at 540 nm (Miranda et al., 2001). Nitrite is usually present in water at a concentration lower than 0.01 mg/L.

The method has been extended to include nitrate by incorporation of a reduction step that converts the nitrate quantitatively to nitrite. The nitrate concentration is found by subtracting the background nitrite from the total nitrate and nitrite obtained after the reduction step. A number of versions of the method have been reported, including the use of vanadium chloride (Wang et al., 2016) or cadmium columns (Gal et al., 2004; Hydes and Hill, 1985; Wood et al., 1967) for the nitrate reduction. The addition of this reduction step makes nitrate detection more complex as a more sophisticated fluidic system is required, further increasing the cost of implementation in an autonomous platform. However, work has recently been carried out to reduce the complexity of these methods. For example, Wang et al. (Wang et al., 2016) presented the use of vanadium chloride (VCl_3) as the reductant instead of the well-known cadmium column for reducing nitrate to nitrite. The work employed an automated flow

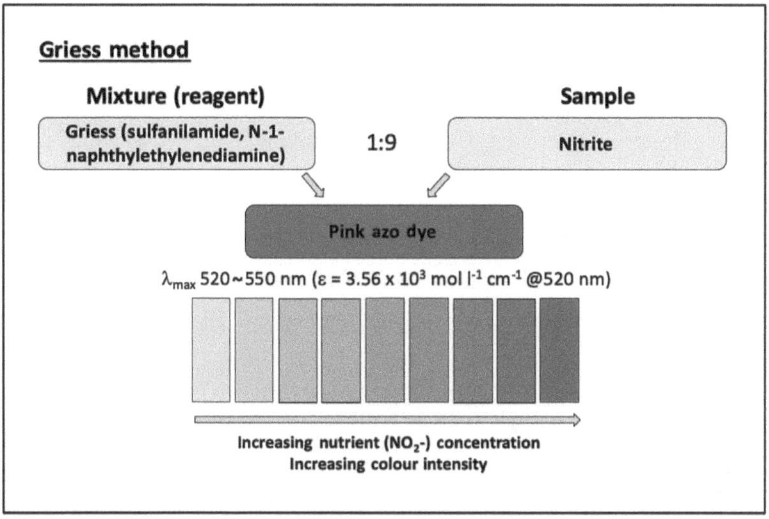

Fig. 7 Conceptual reaction schematic of the Griess assay for colorimetric detection of nitrite (molar extinction coefficient (Sandu et al., 2014)).

injection analysis system using a single analytical solution comprising VCl_3 mixed with the Griess reagent. This method was favored as it removed the need for the use of the toxic cadmium column, and this mixed reagent aimed to simply the fluidic system by combining both the reducing VCl_3 and the reactive Griess reagent into one solution.

7.5 Ammonia colorimetry

The most widely used colorimetric method for the direct determination of ammonia is the Berthelot method. This method has a high sensitivity (LOD at μg/l levels) and an intense color production at 630 nm (Sequeira et al., 2002). The original Berthelot method was based on the formation of a green color when ammonia, phenol, and hypochlorite react (Searle, 1984). In recent years, sodium salicylate has been used in place of the highly toxic phenol. The salicylate version of the method has been applied to environmental sensing using both fresh and marine water samples (Absalan et al., 2004).

In the presence of sodium hydroxide, the Berthelot reagent reacts with ammonia to produce a green color at 660 nm (salicylate). When ammonia forms a monochloramine in the presence of hypochlorite, it is then converted to 5-aminosalicylate by the salicylate, which is then oxidized to produce the blue indophenol complex (Cogan et al., 2014; Sequeira et al., 2002).

The Berthelot method traditionally utilized a three-step process in which three reagents were added sequentially with specific wait times between each step to allow the reaction to take place. To improve the sensitivity of the color development, the solutions were also heated up to 45 °C (Daridon et al., 2001). Daridon et al. (Daridon et al., 2001) successfully implemented the method in a microfluidic system, which compared well to standard spectroscopic detection. However, the sequential multistep approach along with the addition of heat made this application difficult to automate and apply to autonomous sensing. Cogan et al. (Cogan et al., 2014) simplified the traditional method by utilizing a premixed reagent that could be added directly to the sample in a 1:1 *v*/v ratio. This less complex method was much simpler to implement in a microfluidic system (Fig. 8).

8 Current and emerging nutrient-sensing systems

8.1 Examples of current sensing systems

A number of commercially available water quality sensors of various configurations are available. Traditionally, these sensors are typically online analyzers requiring mains power and permanent deployment locations where maintenance and calibration can be carried out and have been more suited to industrial applications than remote deployments. Recently, a number of companies have released in-situ-based nutrient analyzers, some examples of which are presented in Fig. 9.

TresCon UNO (see Fig. 9 image 1) is an online water analysis sensor developed by WTW analyzers. It is capable of continuous measurement of ammonia, phosphate,

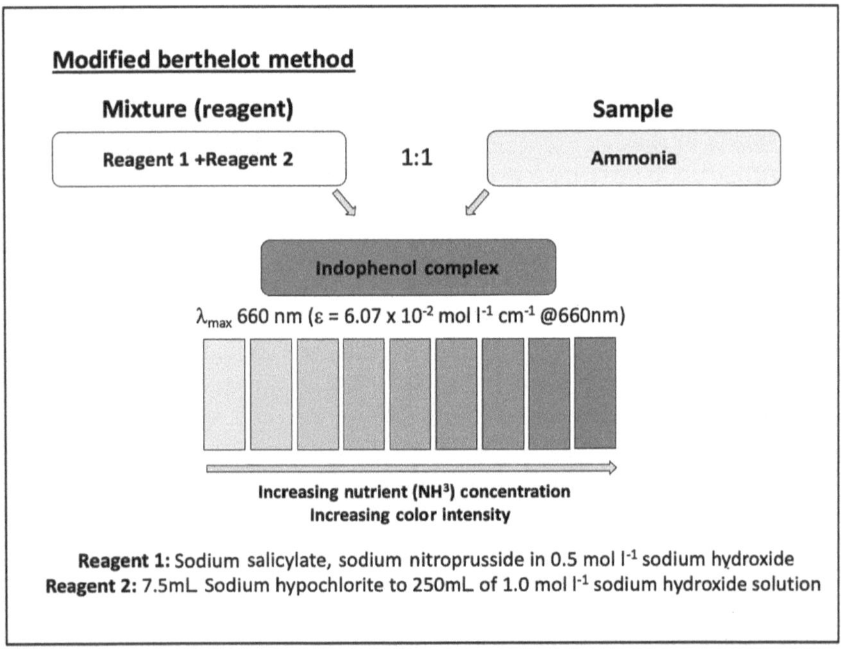

Fig. 8 Conceptual reaction schematic of the Berthelot assay for colorimetric detection of ammonia (molar extinction coefficient (Cogan et al., 2014)).

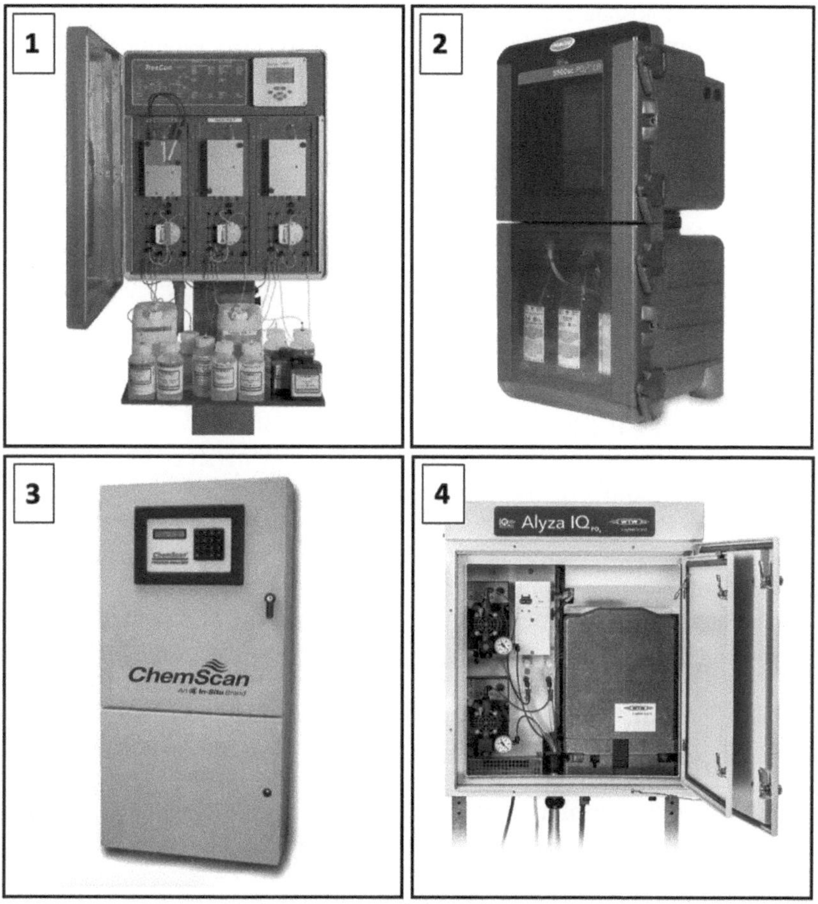

Fig. 9 Commercial water sensing platforms: (1) TresCon UNO (Xylem Analytics, n.d.-a) credit Xylem Analytics Germany, (2) Hach 5500 SC (5500 sc Ammonia Monochloramine Analyzer, n.d.), (3) ChemScan Mini oP (Ortho Phosphorus Analyzer, n.d.), and (4) Alyza IQ (Xylem Analytics, n.d.-b).
Credit Xylem Analytics Germany.

and nitrite or nitrate. It is developed as a modular system that can be set up to measure single or multiple parameters. The TresCon system relies on a continuous supply of water to operate, which is subsampled to supply the sample modules at a typical flow rate of 2–3 l per hour. The phosphate module measures phosphate photometrically using the vanadomolybdophosphoric acid method (yellow method). The nitrate module measures UV absorbance at two wavelengths to ratiometrically obtain an estimate of nitrate concentration by direct absorbance. The ammonia module uses an ammonia-sensitive electrode to measure gaseous ammonia released when the sample is mixed with a basic reagent (Xylem Analytics, n.d.-a).

Hach 5500 SC phosphate analyzer (see Fig. 9 image 2) is an online water analysis sensor with a 15-min measurement cycle, capable of a 90-day runtime and consuming two liters of reagent during this period. The sensor is based on colorimetric chemistries utilizing two different methods depending on the concentration range of phosphate to be detected. The low-range measurement uses the ascorbic acid method, which is measured at 880 nm and has a range of 0.0004 to 5 mg/L PO_4. The high-range measurement uses the molybdovanadate method (yellow method) measured at 480 nm and has a range of 0.02 to 50 mg/L PO_4 (5500 sc Ammonia Monochloramine Analyzer, n.d.).

ChemScan mini oP XR (see Fig. 9 image 3) is a water analysis sensor for the measurement of phosphate using a combination of wet chemistry and spectrophotometry. The sensor comes with low and extended working ranges. The low range measures between 0.003 and 1 mg/L PO_4-P, and the extended working range measures between 0.3 and 60 mg/L PO_4-P. The sensor utilizes large sample tubing to minimize blockages and ensures that any suspended solids less than 150 mg/L can be flushed through the sensor (Ortho Phosphorus Analyzer, n.d.).

Alyza IQ (see Fig. 9 image 4) is an online water analysis sensor capable of measuring ammonia and phosphate. The sensing platform utilizes colorimetric chemistries and automatically calibrates using one- and two-point calibrations. Ammonia is detected using the Berthelot method. A catalyst along with heat is used to accelerate this reaction. Phosphate is detected using the molybdovanadate method (yellow method). The Alyza IQ Ammonia is equipped with two working ranges, 0.02 to 4 mg/L NH_4-N with a resolution of 0.01 mg/L NH_4-N and 0.1 to 20 mg/L NH_4-N with a resolution of 0.05 mg/L NH_4-N. The Alyza Phosphate can have either a low working range of 0.02 to 15 mg/L PO_4-P with a resolution of 0.01 mg/L PO_4-P or 0.2 to 50 mg/L PO_4-P with a resolution of 0.05 mg/L PO_4-P (Xylem Analytics, n.d.-b).

8.2 Emerging sensing systems

Currently, the most readily available commercial nutrient sensors are based on online platforms powered by mains. These typically favor industrial settings where they can regularly be maintained and calibrated.

In-situ deployments are much more challenging, in terms of system maintenance, power management, and unpredictable factors (e.g., weather, deliberate interference, and accidental damage). Most of these novel in-situ analyzers are based around flow systems that intake sample from the water source (Mills and Fones, 2012), mixing with a reagent that produces an optical response proportional to the concentration of the target analyte, measuring the absorbance in the spectral region coinciding with strong absorbance, communicating the resulting data, and storing the waste. Fig. 10 presents a number of sensing platforms that have been developed or are currently being developed around these principles.

The Ecosens Aquamonitrix (see Fig. 10 image 1) is a portable water monitoring platform utilizing ion chromatography. The device is capable of carrying out near-continuous analysis ($\sim$15 min) of water for nitrate and nitrite as well as using reagentless detection to monitor other parameters, reducing reagent consumption

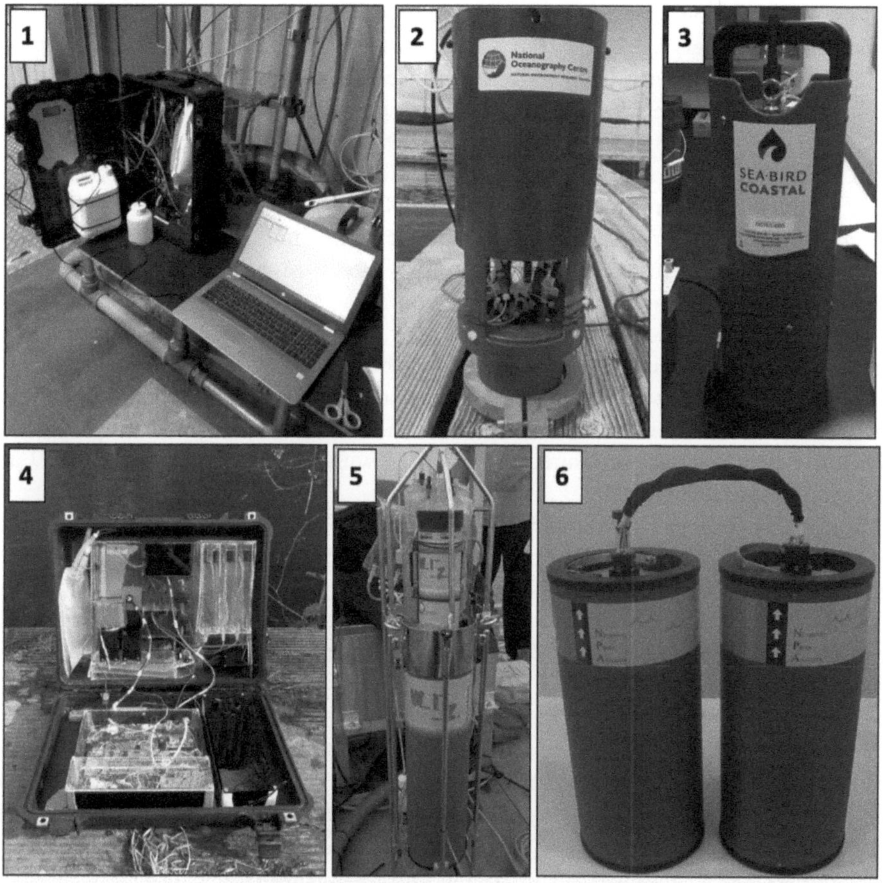

Fig. 10 Prototype sensing platforms for nutrient monitoring: (1) Ecosens Aquamonitrix portable IC system for the determination of nitrite and nitrate (Murray et al., 2019; Ecosens Aquamonitrix, n.d.), (2) National Oceanography Centre, Southampton, phosphate analyzer (Performance Verification Statement for the NOC Phosphate Analyzer, n.d.), (3) Sea-Bird Scientific Hydrocycle PO_4 sensor (Performance Verification Statement for Sea-Bird Scientific HydroCycle-Phosphate Analyzer, n.d.; HydroCycle PO_4, n.d.), (4) Phosphate sensing platform developed by the Adaptive Sensor Group at DCU Insight Centre (Donohoe et al., 2018) (Credit: https://www.insight-centre.org/users/margaret-mccaul), (5) Systea WIZ Probe (WIZ probe, n.d.), and (6) Systea NPA Pro (NPA Pro, n.d.).

(Ecosens Aquamonitrix, n.d.). This portable instrument utilizes a low-cost, UV-absorbent microfluidic detector that incorporates a 235-nm LED (Murray et al., 2019).

The National Oceanography Centre (NOC), Southampton (see Fig. 10 image 2), phosphate analyzer is an autonomous platform based on the vanadomolybdate method and has been applied for the determination of soluble reactive phosphate in seawater. The platform is based on a tinted PMMA microfluidic chip in which liquid flow is delivered via syringe pumps. The linear response has been reported up to 60 μM with a limit of detection of 52 nM (Performance Verification Statement for the NOC Phosphate Analyzer, n.d.).

The Sea-Bird Scientific Hydrocycle PO_4 (see Fig. 10 image 3) is a wet chemical sensor for the in-situ detection of dissolved phosphate. The sensor has a detection limit of 0.0023 mg/L PO_4-P and a nominal range of 0 to 0.3 mg/L PO_4-P. The sensing platform uses optical detection with a wavelength of 870 nm and a path length of 5 cm. The Hydrocycle has a reagent lifetime of 5 months and is capable of taking 1500 measurements between servicing, being capable of measuring up to 4 samples per hour (Performance Verification Statement for Sea-Bird Scientific HydroCycle-Phosphate Analyzer, n.d.).

At the Insight Centre, Dublin City University, (see Fig. 10 image 4), we have developed a phosphate-sensing platform using a combination of microfluidics, colorimetric reagent chemistries (vanadomolybdate method) and low-cost LED-based optical detection. The platform has been tested both in laboratory and in field deployments. The linear response was measured up to 50 μM with a limit of detection of 0.09 μM PO_4^{3-} (Donohoe et al., 2018).

The Systea WIZ Probe (see Fig. 10 image 5) is an in-situ probe that can be used to measure up to four nutrients in surface or sea water. It is capable of measuring ammonia ($N-NH_3$), orthophosphate ($P\text{-}PO_4$), nitrate + nitrite ($N\text{-}(NO_3 + NO_2)$), and nitrite ($N\text{-}NO_2$). The process is carried out using spectrophotometric wet chemistries for phosphate, nitrite, and nitrate and using a fluorometric method for ammonia (Bodini et al., 2015).

The Systea NPA Pro (see Fig. 10 image 6) is an in-situ sensor suited for deployment on remote buoys. The design consists of two separate cylinders, the first containing hydraulic and electrical components and the second containing reagents, calibration standards, DI water, and waste bags. The design is split so that it can be balanced with each cylinder placed into the body holes of the buoy. Similar to the WIZ Probe, the NPA Pro is capable of measuring up to four nutrients. Ammonia is measured using the Berthelot method, phosphate using the molybdenum blue ascorbic acid method, and nitrate/nitrite using cadmium column and Griess method as described earlier (Bodini et al., 2015).

9 Challenges for autonomous in-situ sensing systems

There are a number of limitations to the current real-time autonomous sensing systems, which adds complexity to design and prevent wide-scale adoption of these platforms. In order to function and have wide-scale adoption, issues with robustness, power, detection methods, and fouling must first be addressed. As mentioned previously in Section 3, as the complexity of a sensor increases, the cost of the sensor will also increase and the scalability in terms of deployments will decrease. The cost of the sensing platform relates to not only the initial purchase cost but also its running costs. It is important for sensors to be easy to install, maintain, and use. Software and firmware must be intuitive to use, enabling operators to access, visualize, and analyze data remotely in real time (Schierenbeck and Smith, 2017). The deployment of sensors into the environment exposes them to harsh conditions. The sensor must therefore be capable of withstanding exposure to rain, ice, dust, and variations in temperature and internal humidity. Power consumption remains a major limitation for in-situ sensing if sensing platforms are to be deployed in areas without a local power source. Energy-harvesting options such as solar panels are available, but they can be expensive and as a result can add significantly to the unit cost of the sensor (Knight et al., 2008). The return of investment for solar panels may be in the order of years, which may be longer than the planned lifetime of the deployment. Battery sources still remain the most common method of power for autonomous sensors. Major international research efforts driven by the e-car industry will advance this technology rapidly in the future (Mahmoudzadeh Andwari et al., 2017).

Platform chemistries can also be a considerable limitation for in-situ sensing. Reagent stability is critically important for environmental monitoring due to the need for platforms if they are to be successfully deployed for extended periods of time (Diamond et al., 2012) (at least several months). Reagent-based methods are dependent on the time for the development of a colored product before a measurement can be recorded, and this is highly temperature dependent (Bastian, 1953; Chen et al., 2015; Kanno et al., 2016; Okamura et al., 2014). The time for the colored product to be developed determines the sampling frequency for the sensing platform. The reaction time can be reduced through the use of heat, but the integration of in-line sample heaters affects power draw and lifetime of the battery (Miralles et al., 2013). This along with other factors such as reagent lifetime and reagent consumption rate limits the servicing interval. Careful choice of reagent and calibration standards or extending their lifetime is key to maximizing service intervals and extending deployment periods, which in turn can dramatically reduce the cost of ownership and enhance scalability.

9.1 Sensing platform fouling

Fouling remains one of the largest and almost universal challenges to in-situ sensing within the environment. From the moment a sensing platform is immersed in water, it will begin to be affected by fouling, which disrupts the performance of a sensing platform (Delauney et al., 2010). Fouling can be classified into a number of different types (e.g., colloidal, organic, and biofouling) based on the nature of the effect. Each classification of fouling has its own implications and can have unique remediation or preventative measures when it comes to maintaining sensor performance. Colloidal foulants range in size from 1 nm to 1 μm (Tang et al., 2011). Colloidal fouling occurs when a cake layer is accumulated on the surface or within the pores of a membrane (e.g., prefiltration unit). Clay minerals, colloidal silica, organic colloids, and suspended matter all contribute to colloidal fouling. Organic foulants are caused by substances that are dissolved within the water column and tend to stick to the surface. Common organic foulants include macromolecules, proteins, and oils (Aoustin, 2001).

Biofouling is the formation of living microorganisms or plant life on a surface in contact with water. Biofouling affects the hydraulic performance of sensor inlets making it hard to intake environmental samples through prefilters (Brandt et al., 2017). It is caused by aquatic organisms that can form colonies that cause biofouling due to the interactions of microorganisms with the membrane and dissolved substances within the water column. Biofilms form when free-floating organisms interact and attach with a surface. Initial interactions are usually through reversible, weak van der Walls forces that over time become more permanent due to cell adhesion (Nguyen et al., 2012). The initial attachment of the microbial cells to the surface is known as the conditioning step. A conditioning layer is formed soon after submersion in the water (hours) and leads to the gradual build-up of a thicker biofilm layer over time.

Many approaches have been suggested to combat fouling, but none seem to be universally effective (Manov et al., 2004). Some have been investigated as a means to extend sensor deployment lifetimes, including physical brushes or wipers (shown in Fig. 11 image 4), UV irradiation (Song et al., 2016), or copper/antifouling coatings (Performance Verification Statement for the NOC Phosphate Analyzer, n.d.). Biofouling accumulation drastically affects sensor surfaces, leading to device failure, requiring regular cleaning and replacement of prefilters and other system components. Biofouling does not occur so readily internally within fluidic systems used to implement the reagent-based methods described earlier, due to the relatively harsh chemical conditions required to drive the reagent/sample reaction (e.g., extremely acidic or alkaline conditions and redox chemistries). Walker et al. (Walker, 2015) carried out an investigation of the effects of common reagents used in microfluidic devices including the Griess assay for nitrite on the formation of biofilms by mixed marine communities. It was shown that Griess and other chemical reagents are capable of stopping biofilm formation before they can attach to internal surfaces.

10 Remote sensing

Remote sensing involves the use of satellites or aerial flyovers by aircraft or drones to obtain information about areas from a distance using attenuation or reflection to carry out surface measurements (McCaul et al., 2016). Remote sensing is capable of giving high spatial information about a water body such as sea surface temperatures or color (Andres et al., 2018). Satellite remote sensing in environmental monitoring is capable of providing multispectral and hyperspectral images of continuously increasing spatial accuracy. Satellite remote sensing/earth observation is a growing area with ever-increasing resolution and frequency. The use of this technology allows for the observation of environmental changes at inaccessible locations remotely. Using the information provided by remote sensors such as the EU Copernicus program, Sentinel programs, Landsat 8, and SERVIR allows for changes in the environment to be mapped. Due to its ability to rapidly gather information, the applications of satellite remote sensing and its increasing datasets have allowed it to become a vital part of environmental resource monitoring and management (Chassot et al., 2011; Li et al., 2020; Melesse et al., 2007).

Chen et al. (Chen et al., 2007) presented the use of remote sensing in support of the European Union's Water Framework Directive using the Gulf of Finland as a case

Fig. 11 Biofouling accumulation on deployed water-sensing platforms: (1) Sea-bird Scientific Hydrocycle -PO_4 prior to and following a 32-day field test in the Maumee River (Performance Verification Statement for Sea-Bird Scientific HydroCycle-Phosphate Analyzer, n.d.), (2) prototype nutrient analyzer removed after 3-week deployment, (3) prefilter on the inlet of a nutrient-sensing platform deployed in the marine environment after 1 week, (4) Fouled YSI Sonde after deployment with clean sensing surfaces (Oct 20, 2015) (Image Courtesy: YSI, a Xylem brand), (5) Fluorometer after 30 days in Helgoland (Germany) (Delauney et al., 2010), and (6) NOC sensing platform prior to and following a 32-day field test in the Maumee River (Performance Verification Statement for the NOC Phosphate Analyzer, n.d.).

study. The study focused on using remote sensing to measure the spatial distribution of nitrate concentrations in surface waters. The spatial distribution of this water quality information was produced using empirical algorithms from satellite-based observations, which were verified using in-situ measurements. Satellite remote sensing is capable of providing globally scaled data at an impressive spatial resolution (30 m for Landsat to 1Km for MODIS) (Schaeffer et al., 2013). As the technology relies on satellite or aerial earth imaging, it functions best on clear days where cloud cover is not an issue (Wijedasa et al., 2012; Marshall et al., 1994). In the case of the abovementioned study by Chen et al., the majority of their data set from April to May was cloudy or partly cloudy. For water quality measurements, optical-based observations can be affected by the radiation from direct reflectance from the water surface, but this can be compensated for during data processing (Knight et al., 2008). Remote sensing is limited only to surface layers and requires ground truthing to validate the model used for data processing (Hay, 2000).

The combination of in-situ and remote sensing is a powerful tool for water quality monitoring (Schaeffer et al., 2013). In-situ and remote sensing are complementary approaches, as in-situ measurements are capable of providing validated high-frequency measurements of the local concentration of specific analytes, while remote sensing is capable of providing more generalized data at the global scale (Sputh et al., 2008; Devlin et al., 2013). This complementary approach can provide the ability to actively monitor water processes at a global scale and at a resolution that enables highly localized behavior to be visualized and extended to much larger area (NPA Pro, n.d.). This allows for modeling and understanding of how local events are contributing to the global environment. Tintoré et al. (Tintoré et al., 2019) highlight the need for further cooperation between in-situ and remote sensing as there is little connection at present between the research communities working on satellite remote sensing and in-situ sensing. This lack of a connection is hindering the quality of the data that can be produced from remote sensing. Currently, satellite observations are capable of covering the entire Mediterranean but not at a high enough resolution to capture fine-scale processes that characterize the high temporal and spatial variability of the basin. Further cooperation between the two sensing types would improve the range of parameters that can be observed and carried out remotely, with in-situ ground-truthing improving modeling for satellite sensing.

Albeit at a small scale, the benefits of combining in-situ and remote sensing were illustrated in an investigation to track Marine/Freshwater Mixing Dynamics within Kinvara Bay, Galway, Ireland. McCaul et al. (McCaul et al., 2016) highlighted the advantages of combined sensing where dynamic processes could be directly visualized at a large scale with a resolution that allowed highly localized behavior to be tracked. This work generated large and varied forms of data encompassing in-situ temperature, salinity and phosphate measurements, together with flyover IR thermal imaging, coordinated for temporal and spatial coverage and satellite sea surface temperature measurements. The correlation of this data suggested that it may be possible to track the dynamics of nutrient freshwater/marine mixing in bays using temperature and in-situ ground truthing. In this study, the nutrients entered the bay primarily through freshwater springs, which were sufficiently colder than the warmer sea water to enable tracking of the freshwater movement in the bay by satellite sea surface temperature measurements. As the waterbody (fresh/sea) composition was largely conserved, at least at the surface, the suggestion was the nutrient movements would be

Fig. 12 Correlation between temperature, salinity, and phosphate measurements taken using combination remote and in-situ sensing at Kinvara Bay, Galway, Ireland.
Adapted from McCaul, M., Barland, J., Cleary, J., Cahalane, C., McCarthy, T., Diamond, D., 2016. Combining remote temperature sensing with in-situ sensing to track marine/freshwater mixing dynamics. Sensors 16 (9), 1402. https://doi.org/10.3390/s16091402.

highly correlated with sea-surface temperature (see Fig. 12), reducing the need for dense deployments of autonomous nutrient sensors to track this movement.

This combined sensing strategy will prove to be a powerful tool in the future where complementary methods improve the overall resolution and reliability of a measurement. A combination sensing network where in-situ sensors are taking highly localized data in combination with large area measurement from a remote sensor (satellite, aircraft, and drone) has the potential to drastically change how water monitoring is carried out.

11 Conclusions

Like many other ecosystems, global freshwater, transitional, and marine waters are threatened by the effects of multiple, co-occurring environmental pressures, notably nutrient enrichment and other organic and inorganic pollutants, climate change, and acidification. Real-time monitoring provides reliable in-situ data that will enhance understanding of the processes that drive these changes in water chemistry and therefore strengthen the development of projection models for mitigation measures to reduce impacts seen by contamination such as eutrophication. This chapter highlights the cost and complexity of sensing technology development (and as a result its scalability) depending both on the type of analyte and on the deployment scenario. Physical parameters (e.g., temperature, turbidity, conductivity, and color) are the least complex to track due to the availability of very reliable, low-cost sensors. Physical sensors are typically low power, allowing long-term deployments without the need for servicing or maintenance. However, chemical analytes such as nutrients often require complex detection methodologies and sophisticated sensing platforms to provide accurate in-situ data. Although analytical chemistry techniques have rapidly improved, the cost of in-situ sensors for nutrient detection still remains the prohibiting factor in wide-scale deployments necessary to generate a significant amount of data in order to respond to policies and implement change. Notable advances in electronics, rapid prototyping, microfluidics, and fabrication technologies such as 3D printing have led to the development of new and emerging sensing technologies for the detection of nutrients. However, the global challenge still remains: to develop nutrient sensors at a low cost with specific analytical requirements, coupled with internal calibration, long maintenance intervals, and validation with other analytical techniques. Furthermore, intensive field trials are necessary to make the adoption of these new and emerging technologies feasible for integration into large-scale distributed in-situ sensing networks. Major opportunities still exist for the fusion of in-situ and remote sensing data that is augmented by a comprehensive ground-truthing survey using in-situ sensor networks. Combining the powerful spatial density and wide geographical reach of remote sensing (using drones, flyovers, and satellites) with ground truthing via appropriately located physical and chemical (new and existing) in-situ sensors will generate new data streams that have not been previously accessible. These data have the potential to produce a much more complete and more accurate picture of processes that influence water quality within marine and freshwater bodies.

Acknowledgments

The authors would like to acknowledge T.E. Laboratories Ltd., Co., Carlow, Ireland. Funding from Enterprise Ireland Project Number IP/2016/0502 and from Science Foundation Ireland under Grant SFI/12/RC/2289-P2 (Insight Centre for Data Analytics). We would also like to acknowledge the support under the EU H2020 project Holifab (Grant Number 760927).

References

5500 sc Ammonia Monochloramine Analyzer n.d. Hach USA - Overview | Hach https://www.hach.com/ammonia-monochloramine-analyzers/5500-sc-ammonia-monochloramine-analyzer/family?productCategoryId=35547203805 (accessed August 10, 2019).

Absalan, G., Soleimani, M., Asadi, M., Ahmadi, M.B., 2004. Constructing a new optical sensor for monitoring ammonia in water samples using Bis(Acetylacetoneethylendiamine)

Tributylphosphin cobalt(III) tetraphenylborate complex-coated triacetylcellulose. Anal. Sci. 20 (10), 1433–1436. https://doi.org/10.2116/analsci.20.1433.

Albaladejo, C., Soto, F., Torres, R., Sánchez, P., López, J.A., 2012. A low-cost sensor buoy system for monitoring shallow marine environments. Sensors 12 (7), 9613–9634. https://doi.org/10.3390/s120709613.

Alliance for Coastal Technologies - Nutrient Sensor Challengehttp://www.act-us.info/nutrients-challenge/partcomps/Real_Tech.php n.d. (accessed March 26, 2019).

Anastasova, S., Radu, A., Matzeu, G., Zuliani, C., Mattinen, U., Bobacka, J., Diamond, D., 2012. Disposable solid-contact ion-selective electrodes for environmental monitoring of Lead with ppb limit-of-detection. Electrochim. Acta 73, 93–97. https://doi.org/10.1016/j.electacta.2011.10.089.

Andres, L., Boateng, K., Borja-Vega, C., Thomas, E., 2018. A review of in-situ and remote sensing technologies to monitor water and sanitation interventions. Water 10 (6), 756. https://doi.org/10.3390/w10060756.

Aoustin, E., 2001. Ultrafiltration of natural organic matter. Sep. Purif. Technol. 22–23 (1–2), 63–78. https://doi.org/10.1016/S1383-5866(00)00143-X.

Argo, 2019. Argo Float Data and Metadata from Global Data Assembly Centre (Argo GDAC). https://doi.org/10.17882/42182.

Argo, Part of the Integrated Global Observation Strategy. http://www.argo.ucsd.edu/. (Accessed 30 May 2019).

Ballance, R., Bartram, J. (Eds.), 1998. Water Quality Monitoring: A Practical Guide to the Design and Implementation of Freshwater Quality Studies and Monitoring Programmes. Spon Press, https://doi.org/10.4324/9780203476796.

Bastian, R., 1953. Temperature effects in differential colorimetry. Anal. Chem. 25 (2), 259–262. https://doi.org/10.1021/ac60074a014.

Biological Nitrogen Fixation - An Overview, ScienceDirect Topics. https://www.sciencedirect.com/topics/earth-and-planetary-sciences/biological-nitrogen-fixation. (Accessed 30 May 2019).

Bodini, S., Sanfilippo, L., Savino, E., Moscetta, P., 2015. Automated micro loop flow reactor technology to measure nutrients in coastal water: state of the art and field application. In: OCEANS 2015, Genova. IEEE, Genova, Italy, pp. 1–7, https://doi.org/10.1109/OCEANS-Genova.2015.7271720.

Bol, R., Gruau, G., Mellander, P.-E., Dupas, R., Bechmann, M., Skarbøvik, E., Bieroza, M., Djodjic, F., Glendell, M., Jordan, P., et al., 2018. Challenges of reducing phosphorus based water eutrophication in the agricultural landscapes of Northwest Europe. Front. Mar. Sci. 5, 276. https://doi.org/10.3389/fmars.2018.00276.

Bouabana, A., Şanal, E., Sourkounis, C., 2016. Development of a low cost universal sensor for an accurate measurement of current, voltage and temperature. In: 2016 IEEE International Conference on Renewable Energy Research and Applications (ICRERA), pp. 391–395, https://doi.org/10.1109/ICRERA.2016.7884367.

Bourgeois, W., Romain, A.-C., Nicolas, J., Stuetz, R.M., 2003. The use of sensor arrays for environmental monitoring: interests and limitations. J. Environ. Monit. 5 (6), 852. https://doi.org/10.1039/b307905h.

Bowden, M., 2003. Smaller, Smarter, Faster: The Development and Application of Microfluidic Devices to the Determination of Phosphorus in Natural Waters.

Bowden, M., Diamond, D., 2003. The determination of phosphorus in a microfluidic manifold demonstrating long-term reagent lifetime and chemical stability utilising a colorimetric method. Sensors Actuators B Chem. 90 (1–3), 170–174. https://doi.org/10.1016/S0925-4005(03)00024-8.

Bowden, M., Sequiera, M., Krog, J.P., Gravesen, P., Diamond, D., 2002. Analysis of river water samples utilising a prototype industrial sensing system for phosphorus based on micro-system technology. J. Environ. Monit. 4 (5), 767–771. https://doi.org/10.1039/b200330a.

Brandt, M.J., Johnson, K.M., Elphinston, A.J., Ratnayaka, D.D., 2017. Intakes. In: Brandt, M.J., Johnson, K.M., Elphinston, A.J., Ratnayaka, D.D. (Eds.), Twort's Water Supply, seventh ed. Butterworth-Heinemann, Boston, pp. 205–233, https://doi.org/10.1016/B978-0-08-100025-0.00006-5. Chapter 6.

Bresnahan, P.J., Martz, T.R., Takeshita, Y., Johnson, K.S., LaShomb, M., 2014. Best practices for autonomous measurement of seawater PH with the Honeywell Durafet. Methods Oceanogr. 9, 44–60. https://doi.org/10.1016/j.mio.2014.08.003.

Brogan, J., Crowe, M., Carty, G., 2001. Ireland; Environmental Protection Agency. Developing a National Phosphorus Balance for Agriculture in Ireland: A Discussion Document. Environmental Protection Agency, Johnstown Castle, Co. Wexford.

Chapra, S.C., Boehlert, B., Fant, C., Bierman, V.J., Henderson, J., Mills, D., Mas, D.M.L., Rennels, L., Jantarasami, L., Martinich, J., et al., 2017. Climate change impacts on harmful algal blooms in U.S. freshwaters: a screening-level assessment. Environ. Sci. Technol. 51 (16), 8933–8943. https://doi.org/10.1021/acs.est.7b01498.

Chassot, E., Bonhommeau, S., Reygondeau, G., Nieto, K., Polovina, J.J., Huret, M., Dulvy, N. K., Demarcq, H., 2011. Satellite remote sensing for an ecosystem approach to fisheries management. ICES J. Mar. Sci. 68, 651–666. https://doi.org/10.1093/icesjms/fsq195.

Chen, Q., Zhang, Y., Hallikainen, M., 2007. Water quality monitoring using remote sensing in support of the EU water framework directive (WFD): a case study in the Gulf of Finland. Environ. Monit. Assess. 124 (1–3), 157–166. https://doi.org/10.1007/s10661-006-9215-8.

Chen, L., Ma, J., Huang, Y., Dai, M., Li, X., 2015. Optimization of a colorimetric method to determine trace urea in seawater. Limnol. Oceanogr. Methods 13 (6), 303–311. https://doi.org/10.1002/lom3.10026.

Chen, X., Zhou, G., Mao, S., Chen, J., 2018. Rapid detection of nutrients with electronic sensors: a review. Environ. Sci. Nano 5 (4), 837–862. https://doi.org/10.1039/C7EN01160A.

Cleary, J., Slater, C., McGraw, C., Diamond, D., 2008. An autonomous microfluidic sensor for phosphate: on-site analysis of treated wastewater. IEEE Sensors J. 8 (5), 508–515. https://doi.org/10.1109/JSEN.2008.918259.

Clements, A.L., Griswold, W.G., Rs, A., Johnston, J.E., Herting, M.M., Thorson, J., Collier-Oxandale, A., Hannigan, M., 2017. Low-cost air quality monitoring tools: from research to practice (a workshop summary). Sensors 17 (11), 2478. https://doi.org/10.3390/s17112478.

Cogan, D., Cleary, J., Fay, C., Rickard, A., Jankowski, K., Phelan, T., Bowkett, M., Diamond, D., 2014. The development of an autonomous sensing platform for the monitoring of ammonia in water using a simplified Berthelot method. Anal. Methods 6 (19), 7606–7614. https://doi.org/10.1039/C4AY01359J.

Cogan, D., Fay, C., Boyle, D., Osborne, C., Kent, N., Cleary, J., Diamond, D., 2015. Development of a low cost microfluidic sensor for the direct determination of nitrate using chromotropic acid in natural waters. Anal. Methods 7 (13), 5396–5405. https://doi.org/10.1039/C5AY01357G.

Coleman, S., ter Schiphorst, J., Ben Azouz, A., Bakker, S., Schenning, A.P.H.J., Diamond, D., 2017. Tuning microfluidic flow by pulsed light oscillating spiropyran-based polymer hydrogel valves. Sensors Actuators B Chem. 245, 81–86. https://doi.org/10.1016/j.snb.2017.01.112.

Czugala, M., Ziolkowski, B., Byrne, R., Diamond, D., Benito-Lopez, F., 2011. Materials science: the key to revolutionary breakthroughs in micro-fluidic devices. In: Nano-Opto-Mechanical Systems (NOMS); International Society for Optics and Photonics. 8107.

Daridon, A., Sequeira, M., Pennarun-Thomas, G., Dirac, H., Krog, J.P., Gravesen, P., Lichtenberg, J., Diamond, D., Verpoorte, E., de Rooij, N.F., 2001. Chemical Sensing Using an Integrated Microfluidic System Based on the Berthelot Reaction. 9.

Davidson, K., Gowen, R.J., Harrison, P.J., Fleming, L.E., Hoagland, P., Moschonas, G., 2014. Anthropogenic nutrients and harmful algae in coastal waters. J. Environ. Manag. 146, 206–216. https://doi.org/10.1016/j.jenvman.2014.07.002.

Delaney, C., McCluskey, P., Coleman, S., Whyte, J., Kent, N., Diamond, D., 2017. Precision control of flow rate in microfluidic channels using photoresponsive soft polymer actuators. Lab Chip 17 (11), 2013–2021. https://doi.org/10.1039/C7LC00368D.

Delauney, L., Compère, C., Lehaitre, M., 2010. Biofouling protection for marine environmental sensors. Ocean Sci. 6 (2), 503–511. https://doi.org/10.5194/os-6-503-2010.

Devlin, M.J., da Silva, E.T., Petus, C., Wenger, A., Zeh, D., Tracey, D., Álvarez-Romero, J.G., Brodie, J., 2013. Combining in-situ water quality and remotely sensed data across spatial and temporal scales to measure variability in wet season chlorophyll-a: great barrier reef Lagoon (Queensland, Australia). Ecol. Process. 2 (1), 31. https://doi.org/10.1186/2192-1709-2-31.

Diamond, D., Cleary, J., Maher, D., Kim, J., Lau, K.T., 2011. Autonomous analyser platforms for remote monitoring of water quality. In: IECON 2011 - 37th Annual Conference of the IEEE Industrial Electronics Society. IEEE, Melbourne, VIC, Australia, pp. 3516–3521, https://doi.org/10.1109/IECON.2011.6119878.

Diamond, D., Collins, F., Cleary, J., Zuliani, C., Fay, C., 2012. Distributed environmental monitoring. In: Filippini, D. (Ed.), Autonomous Sensor Networks. Vol. 13. Springer Berlin Heidelberg, Berlin, Heidelberg, pp. 321–363, https://doi.org/10.1007/5346_2012_33.

Dodds, W.K., Bouska, W.W., Eitzmann, J.L., Pilger, T.J., Pitts, K.L., Riley, A.J., Schloesser, J. T., Thornbrugh, D.J., 2009. Eutrophication of U.S. freshwaters: analysis of potential economic damages. Environ. Sci. Technol. 43 (1), 12–19. https://doi.org/10.1021/es801217q.

Donohoe, A., Lacour, G., McCluskey, P., Diamond, D., McCaul, M., 2018. Development of a cost-effective sensing platform for monitoring phosphate in natural waters. Chem. Aust. 6 (4), 57. https://doi.org/10.3390/chemosensors6040057.

DR900 Multiparameter Portable Colorimeter, Hach USA - Overview. https://www.hach.com/dr900-multiparameter-portable-colorimeter/product?id=15684103251. (Accessed 4 June 2019).

Duan, W., He, B., Takara, K., Luo, P., Nover, D., Sahu, N., Yamashiki, Y., 2013a. Spatiotemporal evaluation of water quality incidents in Japan between 1996 and 2007. Chemosphere 93 (6), 946–953. https://doi.org/10.1016/j.chemosphere.2013.05.060.

Duan, W., Takara, K., He, B., Luo, P., Nover, D., Yamashiki, Y., 2013b. Spatial and temporal trends in estimates of nutrient and suspended sediment loads in the Ishikari River, Japan, 1985 to 2010. Sci. Total Environ. 461–462, 499–508. https://doi.org/10.1016/j.scitotenv.2013.05.022.

Duffy, G., Regan, F., 2017. Recent developments in sensing methods for eutrophying nutrients with a focus on automation for environmental applications. Analyst 142 (23), 4355–4372. https://doi.org/10.1039/C7AN00840F.

Duffy, G., Maguire, I., Heery, B., Nwankire, C., Ducrée, J., Regan, F., 2017. PhosphaSense: a fully integrated, portable lab-on-a-disc device for phosphate determination in water. Sensors Actuators B Chem. 246, 1085–1091. https://doi.org/10.1016/j.snb.2016.12.040.

Earthwatch FreshWater Watch n.d. https://freshwaterwatch.thewaterhub.org/ (accessed September 3, 2019)).

Ecosens Aquamonitrix n.d. https://www.ecosensaquamonitrix.eu/ (accessed August 10, 2019).

Environmental Protection Agency. Water Monitoring and Assessment - Environmental Protection Agency (EPA https://www.epa.ie/water/wm/ (accessed May 30, 2019).

Fay, C., Anastasova, S., Slater, C., Buda, S.T., Shepherd, R., Corcoran, B., O'Connor, N.E., Wallace, G.G., Radu, A., Diamond, D., 2011. Wireless ion-selective electrode autonomous sensing system. IEEE Sensors J. 11 (10), 2374–2382. https://doi.org/10.1109/JSEN.2011.2122331.

Frank, C., Schroeder, F., Ebinghaus, R., Ruck, W., 2006. Using sequential injection analysis for fast determination of phosphate in coastal waters. Talanta 70 (3), 513–517. https://doi.org/10.1016/j.talanta.2005.12.055.

Gal, C., Frenzel, W., Möller, J., 2004. Re-examination of the cadmium reduction method and optimisation of conditions for the determination of nitrate by flow injection analysis. Microchim. Acta 146 (2), 155–164. https://doi.org/10.1007/s00604-004-0193-7.

Gimbert, L.J., Haygarth, P.M., Worsfold, P.J., 2007. Determination of nanomolar concentrations of phosphate in natural waters using flow injection with a long path length liquid waveguide capillary cell and solid-state spectrophotometric detection. Talanta 71 (4), 1624–1628. https://doi.org/10.1016/j.talanta.2006.07.044.

Goldman, J., Ramanathan, N., Ambrose, R.F, Caron, D., Estrin, D., Fisher, J., et al., 2007. Distributed sensing systems for water quality assessment and management. Center for Embedded Network Sensing, UCLA. URL https://escholarship.org/uc/item/5v7619xw/. (accessed 05.30.19).

Grand, M.M., Clinton-Bailey, G.S., Beaton, A.D., Schaap, A.M., Johengen, T.H., Tamburri, M. N., Connelly, D.P., Mowlem, M.C., Achterberg, E.P., 2017. A lab-on-Chip phosphate analyzer for long-term in situ monitoring at fixed observatories: optimization and performance evaluation in estuarine and oligotrophic coastal waters. Front. Mar. Sci. 4. https://doi.org/10.3389/fmars.2017.00255.

Hay, S.I., 2000. An overview of remote sensing and geodesy for epidemiology and public health application. Adv. Parasitol. 47, 1–35.

Heisler, J., Glibert, P., Burkholder, J., Anderson, D., Cochlan, W., Dennison, W., Gobler, C., Dortch, Q., Heil, C., Humphries, E., et al., 2008. Eutrophication and harmful algal blooms: a scientific consensus. Harmful Algae 8 (1), 3–13. https://doi.org/10.1016/j.hal.2008.08.006.

Herfort, L., Seaton, C., Wilkin, M., Roman, B., Preston, C.M., Marin, R., Seitz, K., Smith, M. W., Haynes, V., Scholin, C.A., et al., 2016. Use of continuous, real-time observations and model simulations to achieve autonomous, adaptive sampling of microbial processes with a robotic sampler: adaptive sampling in an estuary. Limnol. Oceanogr. Methods 14 (1), 50–67. https://doi.org/10.1002/lom3.10069.

Ho, C.K., Robinson, A., Miller, D.R., Davis, M.J., 2005. Overview of sensors and needs for environmental monitoring. Sensors 5 (2), 4–37.

Hou, D., He, J., Lü, C., Sun, Y., Zhang, F., Otgonbayar, K., 2013. Effects of environmental factors on nutrients release at sediment-water interface and assessment of trophic status for a typical shallow lake, Northwest China. Sci. World J. 2013, 1–16. https://doi.org/10.1155/2013/716342.

Gordon, K., 2016. How chemical monitoring of our rivers can help us understand their story – why, what, where and when. Catchments. ie. URL https://www.catchments.ie/chemical-monitoring-rivers-can-help-us-understand-story/. (accessed 05.30.19).

Hunston, E.WaterBlitz Registration Page https://earthwatch.org.uk/get-involved/waterblitz n.d. (accessed September 3, 2019).

Hydes, D.J., Hill, N.C., 1985. Determination of nitrate in seawater: nitrate to nitrite reduction with copper-cadmium alloy. Estuar. Coast. Shelf Sci. 21 (1), 127–130. https://doi.org/10.1016/0272-7714(85)90011-3.

HydroCycle PO_4 n.d. Sea-Bird Scientific - Overview | Sea-Bird https://www.seabird.com/hydrocycle-po/product?id=54721314201 (accessed September 15, 2019)).

International Initiative on Water Quality: Promoting Scientific Research, Knowledge Sharing, 2015. Effective Technology and Policy Approaches to Improve Water Quality for Sustainable Development., p. 26.

Jaishankar, M., Tseten, T., Anbalagan, N., Mathew, B.B., Beeregowda, K.N., 2014. Toxicity, mechanism and health effects of some heavy metals. Interdiscip. Toxicol. 7 (2), 60–72. https://doi.org/10.2478/intox-2014-0009.

Jang, A., Zou, Z., Lee, K.K., Ahn, C.H., Bishop, P.L., 2011. State-of-the-art lab chip sensors for environmental water monitoring. Meas. Sci. Technol. 22 (3). https://doi.org/10.1088/0957-0233/22/3/032001.

Jokerst, J.C., Emory, J.M., Henry, C.S., 2012. Advances in microfluidics for environmental analysis. Analyst 137 (1), 24–34. https://doi.org/10.1039/C1AN15368D.

Jones, D.O.B., Gates, A.R., Huvenne, V.A.I., Phillips, A.B., Bett, B.J., 2019. Autonomous marine environmental monitoring: application in decommissioned oil fields. Sci. Total Environ. 668, 835–853. https://doi.org/10.1016/j.scitotenv.2019.02.310.

Kanno, S., Cuyas, L., Javot, H., Bligny, R., Gout, E., Dartevelle, T., Hanchi, M., Nakanishi, T. M., Thibaud, M.-C., Nussaume, L., 2016. Performance and limitations of phosphate quantification: guidelines for plant biologists. Plant Cell Physiol. 57 (4), 690–706. https://doi.org/10.1093/pcp/pcv208.

Katewa, S.D., Katyare, S.S., 2003. A simplified method for inorganic phosphate determination and its application for phosphate analysis in enzyme assays. Anal. Biochem. 323 (2), 180–187. https://doi.org/10.1016/j.ab.2003.08.024.

Kaushal, S.S., Groffman, P.M., Band, L.E., Elliott, E.M., Shields, C.A., Kendall, C., 2011. Tracking nonpoint source nitrogen pollution in human-impacted watersheds. Environ. Sci. Technol. 45 (19), 8225–8232. https://doi.org/10.1021/es200779e.

Khan, F.A., Ansari, A.A., 2005. Eutrophication: an ecological vision. Bot. Rev. 71 (4), 449–482. https://doi.org/10.1663/0006-8101(2005)071[0449:EAEV]2.0.CO;2.

Knight, C., Davidson, J., Behrens, S., 2008. Energy options for wireless sensor nodes. Sensors 8 (12), 8037–8066. https://doi.org/10.3390/s8128037.

Kröckel, L., Lehmann, H., Wieduwilt, T., Schmidt, M.A., 2014. Fluorescence detection for phosphate monitoring using reverse injection analysis. Talanta 125, 107–113. https://doi.org/10.1016/j.talanta.2014.02.072.

Kruse, P., 2018. Review on water quality sensors. J. Phys. Appl. Phys. 51 (20). https://doi.org/10.1088/1361-6463/aabb93.

Kumar, P., Morawska, L., Martani, C., Biskos, G., Neophytou, M., Di Sabatino, S., Bell, M., Norford, L., Britter, R., 2015. The rise of low-cost sensing for managing air pollution in cities. Environ. Int. 75, 199–205. https://doi.org/10.1016/j.envint.2014.11.019.

Lazarescu, M.T., 2015. Design and field test of a WSN platform prototype for long-term environmental monitoring. Sensors 15 (4), 9481–9518. https://doi.org/10.3390/s150409481.

Li, J., Pei, Y., Zhao, S., Xiao, R., Sang, X., Zhang, C., 2020. A Review of Remote Sensing for Environmental Monitoring in China. Remote Sens. 12, 1130. https://doi.org/10.3390/rs12071130.

Lucan Bridge-23-08-2019-0 n.d. https://freshwaterwatch.thewaterhub.org/sample/lucan-bridge-23-08-2019-0 (accessed September 19, 2019)).

Mahmoudzadeh Andwari, A., Pesiridis, A., Rajoo, S., Martinez-Botas, R., Esfahanian, V., 2017. A review of battery electric vehicle technology and readiness levels. Renew. Sust. Energ. Rev. 78, 414–430. https://doi.org/10.1016/j.rser.2017.03.138.

Manov, D.V., Chang, G.C., Dickey, T.D., 2004. Methods for reducing biofouling of moored optical sensors. J. Atmos. Ocean. Technol. 21 (6), 958–968. https://doi.org/10.1175/1520-0426(2004)021<0958:MFRBOM>2.0.CO;2.

Manz, A., Graber, N., Widmer, H.M., 1990. Miniaturized total chemical analysis systems: a novel concept for chemical sensing. Sensors Actuators B Chem. 1 (1), 244–248. https://doi.org/10.1016/0925-4005(90)80209-I.

Marshall, G.J., Dowdeswell, J.A., Rees, W.G., 1994. The spatial and temporal effect of cloud cover on the acquisition of high quality Landsat imagery in the European Arctic Sector. Remote Sens. Environ. 50 (2), 149–160. https://doi.org/10.1016/0034-4257(94)90041-8.

McCaul, M., Barland, J., Cleary, J., Cahalane, C., McCarthy, T., Diamond, D., 2016. Combining remote temperature sensing with in-situ sensing to track marine/freshwater mixing dynamics. Sensors 16 (9), 1402. https://doi.org/10.3390/s16091402.

Mehrotra, P., 2016. Biosensors and their applications—a review. J. Oral Biol. Craniofacial Res. 6 (2), 153–159. https://doi.org/10.1016/j.jobcr.2015.12.002.

Melesse, A., Weng, Q., Thenkabail, P., Senay, G., 2007. Remote sensing sensors and applications in environmental resources mapping and modelling. Sensors 7, 3209–3241. https://doi.org/10.3390/s7123209.

Mills, G., Fones, G., 2012. A review of in situ methods and sensors for monitoring the marine environment. Sens. Rev. 32 (1), 17–28. https://doi.org/10.1108/02602281211197116.

Miralles, V., Huerre, A., Malloggi, F., Jullien, M.-C., 2013. A review of heating and temperature control in microfluidic systems: techniques and applications. Diagnostics 3 (1), 33–67. https://doi.org/10.3390/diagnostics3010033.

Miranda, K.M., Espey, M.G., Wink, D.A., 2001. A rapid, simple spectrophotometric method for simultaneous detection of nitrate and nitrite. Nitric Oxide Biol. Chem. 5 (1), 62–71. https://doi.org/10.1006/niox.2000.0319.

Morawska, L., Thai, P.K., Liu, X., Asumadu-Sakyi, A., Ayoko, G., Bartonova, A., Bedini, A., Chai, F., Christensen, B., Dunbabin, M., et al., 2018. Applications of low-cost sensing technologies for air quality monitoring and exposure assessment: how far have they gone? Environ. Int. 116, 286–299. https://doi.org/10.1016/j.envint.2018.04.018.

Mota, L., Mota, A., Pezzuto, C., Carvalho, M., Lavorato, M., Coiado, L., Oliveira, E., 2018. Development of a surface temperature sensor to enhance energy efficiency actions in buildings. Sensors 18 (9). https://doi.org/10.3390/s18093046.

Murray, E., Roche, P., Harrington, K., McCaul, M., Moore, B., Morrin, A., Diamond, D., Paull, B.L., 2019. Cost 235 nm ultra-violet light-emitting diode-based absorbance detector for application in a portable ion chromatography system for nitrite and nitrate monitoring. J. Chromatogr. A. https://doi.org/10.1016/j.chroma.2019.05.036. S0021967319305424.

Nguyen, T., Roddick, F., Fan, L., 2012. Biofouling of water treatment membranes: a review of the underlying causes, monitoring techniques and control measures. Membranes 2 (4), 804–840. https://doi.org/10.3390/membranes2040804.

Nigam, V.K., Shukla, P., 2015. Enzyme based biosensors for detection of environmental pollutants-a review. J. Microbiol. Biotechnol. 25 (11), 1773–1781. https://doi.org/10.4014/jmb.1504.04010.

Nordin, R.N., 2009. Water Quality Guidelines for Nitrogen (Nitrate, Nitrite, and Ammonia)., p. 29.

NPA Pro n.d. http://www.systea.it/index.php?option=com_k2&view=item&layout=item&id=221&Itemid=176&lang=en (accessed September 20, 2019)).

Nutrient Pollution - An Overview n.d.| ScienceDirect Topics https://www.sciencedirect.com/topics/earth-and-planetary-sciences/nutrient-pollution (accessed May 30, 2019).

Oct 20, P.H., 2015. The EXO Wiped Conductivity & Temperature Sensor Helps Eliminate Biofouling. https://www.ysi.com/ysi-blog/water-blogged-blog/2015/10/the-exo-wiped-conductivity-temperature-sensor-helps-eliminate-biofouling. (Accessed 4 June 2019).

Okamura, K., Kimoto, H., Noguchi, T., Hatta, M., Kawakami, H., Suzue, T., 2014. Colorimetric PH measurement for seawater samples using a three light-emitting diodes detector and a calibration method for temperature dependence. Anal. Sci. Int. J. Jpn. Soc. Anal. Chem. 30 (12), 1135–1141. https://doi.org/10.2116/analsci.30.1135.

Ortho Phosphorus Analyzer n.d. https://www.chemscan.com/chemscan-uv-4100-process-analyzer.html (accessed August 10, 2019).

Pellerin, B.A., Stauffer, B.A., Young, D.A., Sullivan, D.J., Bricker, S.B., Walbridge, M.R., Clyde, G.A., Shaw, D.M., 2016. Emerging tools for continuous nutrient monitoring networks: sensors advancing science and water resources protection. J. Am. Water Resour. Assoc. 52 (4), 993–1008. https://doi.org/10.1111/1752-1688.12386.

Performance Verification Statement for Sea-Bird Scientific HydroCycle-Phosphate Analyzer n.d. http://www.act-us.info/Download/Evaluations/NextGenNutrient/SeaBird/ (accessed September 15, 2019)).

Performance Verification Statement for the NOC Nitrate Analyzer n.d. http://www.act-us.info/Download/Evaluations/NextGenNutrient/NOCNitrate/#1 (accessed September 15, 2019)).

Performance Verification Statement for the NOC Phosphate Analyzer n.d. http://www.act-us.info/Download/Evaluations/NextGenNutrient/NOCPO4/#16 (accessed Jun 4, 2019)).

Performance Verification Statement for the Systea WIZ Probe Nitrate Analyzer n.d. http://www.act-us.info/Download/Evaluations/NextGenNutrient/SysteaNitrate/ (accessed September 21, 2019)).

Performance Verification Statement for the Systea WIZ Probe Phosphate Analyzer. http://www.act-us.info/Download/Evaluations/NextGenNutrient/SysteaP04/. (Accessed 21 September 2019).

Phosphate - An Overview n.d. | ScienceDirect Topics https://www.sciencedirect.com/topics/medicine-and-dentistry/phosphate (accessed May 30, 2019).

Popoola, O.A.M., Carruthers, D., Lad, C., Bright, V.B., Mead, M.I., Stettler, M.E.J., Saffell, J.R., Jones, R.L., 2018. Use of networks of low cost air quality sensors to quantify air quality in urban settings. Atmos. Environ. 194, 58–70. https://doi.org/10.1016/j.atmosenv.2018.09.030.

Pretty, J.N., Mason, C.F., Nedwell, D.B., Hine, R.E., Leaf, S., Dils, R., 2003. Environmental costs of freshwater eutrophication in England and Wales. Environ. Sci. Technol. 37 (2), 201–208. https://doi.org/10.1021/es020793k.

Prien, R.D., 2007. The future of chemical in situ sensors. Mar. Chem. 107 (3), 422–432. https://doi.org/10.1016/j.marchem.2007.01.014.

Radu, T., 2009. Ireland; Environmental Protection Agency; EPA STRIVE Programme. Analytical Devices for Autonomous Monitoring of the Environment. Environmental Protection Agency, Johnstown Castle, Co. Wexford.

Sandu, M., Lupascu, T., Tarita, A., Goreacioc, T., Turcan, S., Mosanu, E., 2014. Method for nitrate determination in water in the presence of nitrite. Chem. J. Mold. 9 (2), 8–13. https://doi.org/10.19261/cjm.2014.09(2).01.

Schaeffer, B.A., Schaeffer, K.G., Keith, D., Lunetta, R.S., Conmy, R., Gould, R.W., 2013. Barriers to adopting satellite remote sensing for water quality management. Int. J. Remote Sens. 34 (21), 7534–7544. https://doi.org/10.1080/01431161.2013.823524.

Schierenbeck, T.M., Smith, M.C., 2017. Path to impact for autonomous field deployable chemical sensors: a case study of in situ nitrite sensors. Environ. Sci. Technol. 51 (9), 4755–4771. https://doi.org/10.1021/acs.est.6b06171.

Searle, P.L., 1984. The Berthelot or indophenol reaction and its use in the analytical chemistry of nitrogen. A review. Analyst 109 (5), 549. https://doi.org/10.1039/an9840900549.

Sequeira, M., 2002. Towards autonomous environmental monitoring systems. Talanta 56 (2), 355–363. https://doi.org/10.1016/S0039-9140(01)00601-4.

Sequeira, M., Diamond, D., Daridon, A., Lichtenberg, J., Verpoorte, S., Rooij, N.F.d., 2002. Progress in the realisation of an autonomous environmental monitoring device for ammonia. TrAC Trends Anal. Chem. 21 (12), 816–827. https://doi.org/10.1016/S0165-9936(02)01205-0.

Singh, R.; Singh, S. P. Development of a Low Cost Wireless Temperature Monitoring System for Industrial & Research Application. 7.

Smith, V.H., 2003. Eutrophication of freshwater and coastal marine ecosystems a global problem. Environ. Sci. Pollut. Res. 10 (2), 126–139. https://doi.org/10.1065/espr2002.12.142.

Song, K., Mohseni, M., Taghipour, F., 2016. Application of ultraviolet light-emitting diodes (UV-LEDs) for water disinfection: a review. Water Res. 94, 341–349. https://doi.org/10.1016/j.watres.2016.03.003.

Sputh, B.H.C., Faust, O., Pettersson, L.H., Hamre, T., Vitulli, D., Allen, A.R., Spracklen, T., 2008. Integration of in-situ and remote sensing data for water risk management. In: Presented at the Proc. iEMSs 4th Biennial Meeting – Int. Congress on Environmental Modelling and Software: Integrating Sciences and Information Technology for Environmental Assessment and Decision Making, iEMSs 2008, pp. 486–493.

Swinehart, D.F., 1962. The Beer-Lambert Law. J. Chem. Educ. 39 (7), 333. https://doi.org/10.1021/ed039p333.

Tang, C.Y., Chong, T.H., Fane, A.G., 2011. Colloidal interactions and fouling of NF and RO membranes: a review. Adv. Colloid Interf. Sci. 164 (1–2), 126–143. https://doi.org/10.1016/j.cis.2010.10.007.

Tchounwou, P.B., Yedjou, C.G., Patlolla, A.K., Sutton, D.J., 2012. Heavy metals toxicity and the environment. EXS 101, 133–164. https://doi.org/10.1007/978-3-7643-8340-4_6.

The Argo Program Observing the Global Ocean with Profiling Floats n.d. https://www.researchgate.net/publication/33549908_The_Argo_Program_Observing_the_Global_Ocean_with_Profiling_Floats (accessed May 30, 2019).

Thierry, V., Bittig, H., Gilbert, D., Kobayashi, T., Kanako, S., Schmid, C., 2018. Processing Argo Oxygen Data at the DAC Level; Ifremer. https://doi.org/10.13155/39795.

Tintoré, J., Pinardi, N., Álvarez-Fanjul, E., Aguiar, E., Álvarez-Berastegui, D., Bajo, M., Balbin, R., Bozzano, R., Nardelli, B.B., Cardin, V., et al., 2019. Challenges for sustained observing and forecasting systems in the Mediterranean Sea. Front. Mar. Sci. 6, 568. https://doi.org/10.3389/fmars.2019.00568.

To, Y.S., Randall, C.W., 1977. Evaluation of ascorbic acid method for determination of orthophosphates. J. Water Pollut. Control Fed. 49 (4), 689–692.

Walker, D.I., 2015. Biofouling and its Control for In Situ Lab-on-a-Chip Marine Environmental Sensors., https://doi.org/10.13140/rg.2.1.4275.5040.

Wang, S., Lin, K., Chen, N., Yuan, D., Ma, J., 2016. Automated determination of nitrate plus nitrite in aqueous samples with flow injection analysis using vanadium (III) chloride as reductant. Talanta 146, 744–748. https://doi.org/10.1016/j.talanta.2015.06.031.

Ward, M.H., Jones, R.R., Brender, J.D., de Kok, T.M., Weyer, P.J., Nolan, B.T., Villanueva, C. M., van Breda, S.G., 2018. Drinking water nitrate and human health: an updated review. Int. J. Environ. Res. Public Health 15 (7). https://doi.org/10.3390/ijerph15071557.

Wijedasa, L.S., Sloan, S., Michelakis, D.G., Clements, G.R., 2012. Overcoming limitations with Landsat imagery for mapping of peat swamp forests in Sundaland. Remote Sens. 4 (9), 2595–2618. https://doi.org/10.3390/rs4092595.

Williams, R.J., Johnson, A.C., Smith, J.J.L., Kanda, R., 2003. Steroid estrogens profiles along river stretches arising from sewage treatment works discharges. Environ. Sci. Technol. 37 (9), 1744–1750.

WIZ probe n.d. http://www.systea.it/index.php?option=com_k2&view=item&layout=item&id=223&Itemid=175&lang=en (accessed September 20, 2019)).

Wood, E.D., Armstrong, F.A.J., Richardsl, F.A., 1967. Determination of nitrate in sea water by cadmium-copper reduction to nitrite. J. Mar. Biol. Assoc. UK, 23–31.

Xylem Analytics n.d.-a WTW's Analyzer https://www.xylemanalytics.com/en/products/analyzers (accessed August 10, 2019).

Xylem Analytics n.d.-b Ammonium-Measuring by WTW https://www.xylemanalytics.com/en/parameters/ammonium (accessed August 10, 2019)).

Yang, X., Wu, X., Hao, H., He, Z., 2008. Mechanisms and assessment of water eutrophication. J Zhejiang Univ Sci B 9 (3), 197–209. https://doi.org/10.1631/jzus.B0710626.

Yool, A., Popova, E.E., Anderson, T.R., 2013. MEDUSA-2.0: an intermediate complexity biogeochemical model of the marine carbon cycle for climate change and ocean acidification studies. Geosci. Model Dev. 6 (5), 1767–1811. https://doi.org/10.5194/gmd-6-1767-2013.

Young, D.T., Chapman, L., Muller, C.L., Cai, X.-M., Grimmond, C.S.B., 2014. A low-cost wireless temperature sensor: evaluation for use in environmental monitoring applications. J. Atmos. Ocean. Technol. 31 (4), 938–944. https://doi.org/10.1175/JTECH-D-13-00217.1.

Pervasive sensing approaches for environmental pollution

3

Francesco Pilla[a], *Bidroha Basu*[a], *Salem Gharbia*[b], *and Panagiota Katsikouli*[c]
[a]School of Engineering, University College Dublin, Dublin, Ireland, [b]Department of Civil Engineering & Construction, IT SLigo, Sligo, Ireland, [c]*DTU* COMPUTE Department of Applied Mathematics and Computer Science, Technical University of Denmark, Lyngby, Denmark

1 Introduction

The evergrowing populations and increasing human activities are associated with a major increase in environmental pollution. However, their impact is still poorly understood due to the pervasiveness and spatiotemporal complexity of the phenomenon. Conventional approaches to environmental monitoring are based on networks of sparse measurement stations or in situ human-operated measurements. However, these are prohibitively expensive to capture the spatiotemporal heterogeneity of most of the environmental pollution phenomenon. Current advancements in the wireless sensor network (WSN) technology are radically changing the conventional approach, allowing for the capture of real-time information in a capillary form (Kumar et al., 2015).

A WSN is a self-organized wireless network composed of a large number of spatially distributed sensors to monitor the environmental conditions and transmit the monitored data to the main location via a network. The developments in wireless network technology and miniaturization make it possible to realistically monitor the natural environment and detect environmental contaminants at high spatiotemporal resolutions. It has to be noted, though, that domain knowledge is an essential component of any successful and representative monitoring campaign using WSN, because it is of critical importance to understand the physical environment and deployment in detail before designing and installing any system (Martinez et al., 2004). Sensors that detect pollution in various environmental media (air, water, and soil) can adopt different measurement techniques (Kumar et al., 2011); here are some examples:

- *Chromatography sensors* rely on separation of complex mixtures by percolation through a selectively adsorbing medium, with subsequent detection of compounds of interest.
- *Electrochemical sensors* detect variations in resistance through electrodes because of the electrical current that interacts with chemicals.
- *Mass sensors* measure changes to the mass of the surface of the sensor during interaction with chemicals.
- *Optical sensors* detect changes in visible light or other electromagnetic waves during interactions with chemicals.

Monitoring Environmental Contaminants. https://doi.org/10.1016/B978-0-444-64335-3.00005-0

- *Chemical sensors* measure a wide range of elements and molecules for organic, inorganic, and biochemical molecules in all environmental media (atmosphere, soils, sediments, groundwater, and fresh and marine waters).

The following sections present several case studies of WSNs used to pervasively monitor air, noise, water, and soil pollution. The case studies are presented in the form of examples of deployments, focusing on challenges, opportunities, and recommendations.

2 Air

Air quality in the built-up environment is a critical topic due to its severe impact on public health. It is essential to monitor and control air pollution, especially in and around cities where a majority of the world's population lives and the pollution levels are also typically much higher than those in rural areas. Therefore, it is very crucial to develop, calibrate, and validate spatially distributed tools for real-time air quality monitoring in order to allow for appropriate and timely decisions, taking into account its cost of deployment.

Conventional air quality monitoring approaches involve the deployment of networks of static and sparse high-level monitoring stations. However, these are prohibitively expensive for the capturing of temporospatial heterogeneity and the identification of pollution hot spots, which are required for the development of robust real-time strategies for exposure control (Kumar et al., 2015). This provides a highly accurate description of the pollution at the measurement points, but it lacks the required density to provide a representative description of the spatiotemporal variability of pollution in the complex environment of a city. As such, a sparse network of high-level monitoring stations is complemented with extensive modeling to bridge such a gap. The current progress in developing low-cost microscale sensors is generating a radical shift from the conventional approach allowing for real-time information capture in a capillary form (Kumar et al., 2015; Rai et al., 2017). The deployment of pervasive low-cost sensors in highly significant numbers can help in increasing the spatial and temporal quality of the monitored pollutant levels and also assist in detecting pollution hot spots, in addition to allowing for real-time exposure assessment.

Recently developed air quality sensors use microfabrication techniques and contain microelectromechanical systems made of microfluidic, optical, and nanostructured elements, allowing them to be small, lightweight, and of low cost (Kumar et al., 2015). These newly developed sensors are equipped with circuits that have very low power consumption and electricity-efficient communication kit. These are also accompanied with advanced computing techniques for data handling and a wide choice of software programmed for data handling and visualization. These sensors have elements that respond to variations in physical or chemical properties that are converted to electrical signals by the transducers (Kumar et al., 2015).

In general, for these kits, chemical gas sensors determine the concentration of different gases by quantifying the reactions between the sensing material and target gases, for example, ozone (O_3), carbon monoxide (CO), sulfur dioxide (SO_2), carbon dioxide (CO_2), nitrogen dioxide (NO_2), and volatile organic compounds. Optical gas sensors quantify the adsorption of light by the gaseous species of interest. However, light scattering is utilized for quantifying particle number concentrations that can be translated to any mass fraction, for example, PM_{10} and $PM_{2.5}$ (Kumar et al., 2015; White et al., 2012).

In the past decade, there has been a widespread growth in both academia- and industry-led applications aimed at developing and deploying this new generation of low-cost air quality sensors for WSNs. Many of the research efforts have been focused on identifying appropriate protocols to practically use the developed sensors and networks. Efforts have also been taken to develop data analysis and visualization tools, in addition to the experiments on how pervasive low-cost sensors respond under different environmental conditions. The development and research programs have led to a wide range of low-cost air quality sensors suitable for pervasive applications that are commercially available (Morawska et al., 2018; Rai et al., 2017).

2.1 Examples of deployments

During the past 5 years, many WSN deployments have been conducted for disparate purposes, for example, residents evaluating air quality in their premises or neighborhoods, small networks looking at the concentration of air pollutants at community level, and pervasive sensor networks covering cities or large regions. Some researchers are also deploying pervasive sensor networks to assess their performance for both static-site and mobile measurements (Brienza et al., 2015; Kaufman et al., 2017; Gao et al., 2015; Sun et al., 2016; Jiao et al., 2016).

Jiao et al. (2016) presented the Community Air Sensor Network (CAIRSENSE) project that deployed low-cost, continuous, and commercially available air pollution sensors in the southeastern United States. The deployment monitored NO, NO_2, O_3, CO, SO_2, and PM_{10}. The main observation from this campaign is that sensors have a highly variable performance with weak sensitivity, and they concluded that it was necessary to incorporate collocation with a high-end reference device in any future study.

Brienza et al. (2015) presented *uSense,* which is a low-cost air quality sensor. *uSense* allows for the capture of real-time gas concentrations. The users can monitor their areas, and they can share the collected data via social networking.

Sun et al. (2016) reported the development and the application of next-generation air quality sensors. The application was for the international marathon in Hong Kong. This air monitoring network deployment allowed to conclude that traffic control applied for the marathon was very efficient in decreasing the levels of air pollutants such as NO_2, and CO, while $PM_{2.5}$ as less affected.

Gao et al. (2015) examined the performance of a low-cost pervasive sensor in a highly polluted built-in environment. Some Portable University of Washington Particle (PUWP) sensors were calibrated with optical and gravimetric $PM_{2.5}$ reference stations and then used for this deployment. The main conclusion was that the PUWP sensors could be used to enhance existing $PM_{2.5}$ monitoring networks in order to increase the spatiotemporal resolution of the available $PM_{2.5}$ datasets.

2.2 Challenges, opportunities, and recommendations

The primary challenge related to the use of low-cost sensors for pervasive WSNs for air pollution monitoring is improving their sensitivity, stability, and longevity of operation. Data records from the spatially distributed network of sensors can be transmitted in real time and stored in online databases and platforms, which provides a perfect way for quick dissemination. However, the ownership of the logged data provides both challenges and opportunities, which should be addressed effectively and transparently through GDPR-compliant strategies. Williams et al. (2014) presented the Air Sensor Guidebook developed by US Environmental Protection Agency, which is a precious resource to design the data collection process using a network of sensors. Reece et al. (2017) also provided design considerations for low-cost sensors, based on the abilities of the user, the length of deployment, and the measurement goals of the deployment.

Another challenge is the calibration and validation of low-cost sensors, which are generally affected by environmental factors and thus may affect the transferability of the technologies between different environmental conditions (Rai et al., 2017). Different methods of sensor calibration against reference stations were investigated, which showed the presence of complex nonlinear and cross-sensitive behavior of sensors. Many experimental normalization techniques to solve these complex behaviors, including multilinear regressions, nonlinear multivariate models, and machine learning, have been deployed (Cavaliere et al., 2018; Zimmerman et al., 2018; Cordero et al., 2018; Malings et al., 2019), and the scientific community still works on this challenge using different new computational techniques.

3 Noise

Wireless sensor networks have been excessively used for monitoring and contextually characterizing various environments. When it comes to sound, not only standalone microphones, but also those embedded in smart devices, have been used for activity recognition (Rossi et al., 2013; Lu et al., 2009; Aiello et al., 2016), real-time notification of domestic or road accidents (White et al., 2011; Zigel et al., 2009), and health monitoring (Lu et al., 2012; Hao et al., 2013). Sound has also been used as a proxy to assess human experience of a particular location (Aiello et al., 2016). In urban environments, ambient noise is a major contributing factor not only for distress and annoyance but also for heart-related issues, sleep disorders, and antisocial behavior (European Commission, 2002). Therefore, monitoring noise levels can not only help understand the extent of noise pollution but also help organize urban and rural life in ways that reduce exposure to it. The personal nuisance from noise pollution can be computed as a function of loudness, sharpness, roughness, and the fluctuation strength of the recorded sound levels (Zwicker and Fastl, 2013).

The European Noise Directive has instructed European member states to construct noise maps mainly for the road networks, railways, and airports. As of 2007, cities with population above 250,000 were to estimate the number of citizens exposed to noise levels of 55–75 dB in bands of 5 db and those exposed to noise pollution above 75 dB. Different values were suggested for road networks, railways, and air traffic. As of 2012, cities with a population of at least 100,000 were instructed to be included in the monitoring process.

Noise-monitoring activities using WSNs have also been performed for ad hoc purposes as part of grassroots and citizen science activities aimed at specific issues by leveraging smartphone microphone sensors. Examples of these are the NoiseTube (Maisonneuve et al., 2009a,b) and WideNoise (Becker et al., 2013) applications.

3.1 Examples of deployments

Traditionally, noise monitoring has been done by measuring sound levels at predefined locations for specific periods of time. For example, in Zurich, Switzerland, noise pollution was estimated through manual recordings of traffic noise, by counting cars passing through in intervals of 30 min. Spatiotemporal extrapolation through dedicated commercial noise-propagation models is used to estimate noise levels at nonmonitored areas. Since 2006, researchers and city councils have shifted their attention to nonmanual monitoring of noise, through the use of WSNs (Navarro et al., 2017). In 2006, in Barcelona, Spain, a long-term noise-monitoring sensor network was set up in different areas of the city (Farrés, 2015). A main network comprising nearly 30 sound-level meters and a complementary network of 11 low-cost sound sensors were deployed at two construction sites, partly at fixed locations and partly offering on-demand coverage. In over 8 years of operation, the network collected measurements at nearly 100 distinct locations. The main source of noise was found to be recreational (57%), whereas road noise accounted for 20% of the noise. Dublin City Council has a similar deployment using an Internet of Things solution with networked sound sensors located in 40 locations (Coletta and Kitchin, 2017).

In Zurich (Santini et al., 2008), three different platforms were tested on side roads: (i) Tmote Sky from Moteiv, which was abandoned due to unsatisfactory performance, (ii) Tmote from Moteiv, which requires a sampling rate of 32 kHz, and (iii) their own-built noise-level meter. Variance in the average output voltage and the number of collected samples was observed when the sensors were plugged in a power source as compared to when they were running on their own battery. Low-cost Tmote-Invent nodes and Raspberry Pi sensors were used in experiments monitoring noise from vehicular traffic on a highway near Valencia (Segura-Garcia et al., 2015). The nodes have an estimated lifetime of 78 days, whereas a Raspberry Pi has only 39 h. The authors concluded in favor of the use of the Raspberry Pis due to their lower cost and higher performance, despite their low battery life.

Traditionally deployed noise sensor networks have high performance-and-maintenance costs, coarse-grained coverage of the environment, and natural limitations in reflecting human nuisance. On the other hand, using microphones embedded in smartphones is a ubiquitous alternative that overcomes the aforementioned drawbacks.

Participatory sensing has a very small cost, mainly coming from transmitting and processing the collected data. Even though different smart phones have different calibrations due to their manufacturer specifications, correlation of measurements with those coming from sound-level meters has been shown to be high (D'Hondt et al., 2013). In terms of coverage, the mobility of a smartphone carrier increases the spatiotemporal density of collected samples, compared to those coming from a few deployed sensors. However, in order to avoid a random distribution of the readings, a group of users is needed instead of a single individual (Rana et al., 2010). In that fashion, the silent and noisy areas become prominent beyond uncertainty. Also, participatory sensing naturally offers a realistic measurement of human annoyance caused by noise pollution, as the users are usually asked to give contextual and annoyance rating information for the various monitored areas. Such a task might seem obtrusive; however, volunteers in a participatory sensing experiment (Kanjo, 2010) found the process to be fun and interesting.

NoiseTube is one of the first participatory sensing platforms for noise pollution (Maisonneuve et al., 2009a; Stevens and D'Hondt, 2010), and it was used in a case study in Belgium for the construction of the noise map of a neighborhood (Stevens and D'Hondt, 2010; D'Hondt et al., 2013). Given that the accuracy and reliability of the results depend on the number and distribution of collected measurements, formula (1) presents the number of measurements (M) as a function of the number of days (D), the number of volunteers (n), the collection period in minutes (T), and the rate of measurements per minute (R) for monitoring an area of width (W) and height (H) tessellated in a grid cell size of r meters (D'Hondt et al., 2013).

$$M = (R*n*T*D*r\wedge 2)/(W*H) \tag{1}$$

Another grassroots participatory sensing deployment adopting a citizen science approach is the WideNoise application, which encouraged the local communities around Heathrow Airport to download the app to take noise measurements in the surrounding area (Becker et al., 2013; Atzmueller et al., 2012). This pilot was mostly focused on developing tools to directly involve citizens in environmental monitoring, with the aim to encourage learning and enhance awareness of environmental issues.

3.2 Challenges, opportunities, and recommendations

The reliability of the results in a participatory experiment could be influenced by factors such as wind or discussions. In a use case in Cambridge (Kanjo, 2010), where pedestrians and cyclists collected sound data with their phones, the use of background noise elimination techniques or that of the correction factor was proposed. In terms of power consumption, it is shown (Kanjo, 2010) that the battery usage during collection is only slightly higher compared to the standby mode.

In general, participatory sensing is a cheap and reliable way to construct noise maps, closer to the human experience of noise pollution. Maps constructed from fixed sensor networks or simulations (based on traffic conditions, for example) can benefit from participatory sensing, as the two in unison can give a significantly more accurate result (Hachem et al., 2015).

4 Water

Pollution of surface water has become a major problem over the past few decades worldwide. Currently, approximately 250 million people suffer from water-related diseases, leading to 5–10 million deaths every year (Chung and Yoo, 2015), while approximately one-sixth of the global population does not have access to clean drinking water (Kruse, 2018). Water distribution systems (WDSs), which provide water for drinking, household, industrial and agricultural purposes, are vulnerable to accidental/intentional contamination as they are accessible and unprotected (US EPA, 2003). Recent examples of accidental contamination include nitrobenzene contamination in the Songhua River, China, due to a chemical plant explosion in 2005 and the benzene contamination due to crude oil leak from a petrochemical pipeline at Lanzhou, China, in 2014, both leading to suspension of water supply service for days. The benzene contamination was detected days after the event, justifying the necessity of nearly real-time water quality monitoring of WDSs. Furthermore, pollution monitoring in streams, rivers, lakes, and coastal areas is of paramount importance as over 40% of the world's population resides in the shores of those natural waterbodies (Haji Gholizadeh et al., 2016).

4.1 Examples of deployments

Water pollution in WDSs as well as natural waterbodies can be monitored either by collecting samples from the field and then analyzing the samples in the laboratory or by using a WSN. Field measurement offers high accuracy but is a time-consuming and labor-intensive process. Hence, to analyze water quality for large water areas, complex terrains, remote areas, or large-scale WDSs exhibiting variability and complexity, a WSN is considered to be an effective option. The main advantage of WSNs is that the sensors can be easily deployed at important locations at a considerably low cost. A WSN records data at continuous time and is able to cover a wider region of the WDS or a natural waterbody. The data obtained from a WSN can be used to identify the spatial or temporal variations of water quality, which is essential for comprehensive assessment and management of waterbodies and for the reduction of the social and economic damages caused by water pollution. Since the water quality information can be recorded and transmitted in real time, the WSN is an effective tool for an early warning system (EWS). An efficient EWS should detect the contamination event from normal variations and identify the type of contaminant quickly to support the implementation of remediation attempts (Liu et al., 2015).

WSNs for water monitoring can be broadly categorized into two types: conventional and compound-specific/advanced water quality sensors. The conventional sensors test a set of water quality parameters classified as physical, chemical, and biological parameters. Physical parameters include density, turbidity, color, temperature, conductivity, total dissolved solids, and total suspended solids (TSS); chemical parameters consist of inorganic (pH, dissolved oxygen, acidity/alkalinity, hardness, oxidation-reduction potential, and chlorine) and organic (decaying plant and animal excrement, total organic carbon (TOC), pesticides) compounds; and biological parameters include various organisms from algae and phytoplankton to bacteria and human pathogens (Kruse, 2018). Most of those parameters can be monitored by using a combination of optical sensors, microelectronics mechanical systems, and biosensors as those sensors have high sensitivity, selectivity, insensitivity to electromagnetic interference, and quick response time, leading to the possibility of real-time analysis of data with high confidence. The advanced water quality sensors are based on emerging detection techniques that have the capability to detect low concentrations for a specific component (Jeon et al., 2008).

4.2 Challenges, opportunities, and recommendations

The key problem of WSNs is that it is difficult to integrate all of the different parameter detection techniques in a single system due to the varying nature of water parameters (Bhardwaj et al., 2015). Most WSNs are limited to identifying a single or a small class of pollutants, leading to inefficiency and recognition failure (Liu et al., 2015; Huang et al., 2017; Qin et al., 2018). To overcome this limitation, researchers have used a set of conventional parameters to develop a real-time classification method of pollutants. Water Security Initiative Program in the United States considered pH, turbidity, temperature, conductivity, TOC, and chlorine as the primary set of pollutants to quantify water quality (US EPA, 2005, 2008). Kroll (2006) combined five water pollution data (pH, conductivity, turbidity, chlorine residual, and TOC) obtained from multiple sensors into a single scalar value by using the Hach Homeland Security Technologies approach for detection and contaminant classification. In situations where the value exceeds a prespecified threshold, water was considered to be contaminated and further classifications were performed to identify the cause of the contamination. Yang et al. (Jeffrey Yang et al., 2009) conducted a set of experiments by observing the response time of 11 different pollutants and developed a real-time event adaptive detection, classification, and warning method for event detection and contamination classification. Liu et al. (2015) used Mahalanobis distance measure to identify the contaminant pollutant, whereas Huang et al. (2017) proposed a multi-classification probability output approach by considering multiclassification support vector machine algorithm for classification of pollutants.

Another important aspect of WSNs is that the monitoring can be performed in a remote/harsh environment, such as in a lake area or underwater. In those situations, it is difficult to reconfigure the WSN manually. The network should have the flexibility of self-configuration with smart decision support for specific purposes. Examples of such purposes include saving sensor energy by stopping the monitoring during uninteresting events and autocalibrating the sensor to address the drift over time due to biofouling of the probes (Huang et al., 2015; Leigh et al., 2019). Along with WSNs, remote sensing data obtained from spaceborne (satellite) and airborne (aircraft, helicopter, balloon, boat, etc.) sensors are in use for water quality analysis. These sensors measure the radiation reflected from water surface at various wavelengths, which in turn can be used to detect a set of water quality indicators including turbidity, salinity, temperature, pH, TSS, TOC, chlorophyll-a concentration, and total phosphorus (Haji

Gholizadeh et al., 2016). Recently, nanomaterial sensor platforms using nanowires (Cui et al., 2001) and quantum dots (Han et al., 2001) has gained considerable attention in water quality analysis. The major advantage of these nanomaterial sensors is that they have the ability to detect ultralow concentration of complex pollutants in a very rapid analysis time period (Vikesland, 2018). Testing of the nanomaterial sensor is underway for varying environments.

5 Soil

The diversity and quantity of chemicals released by human activities into the environment have risen dramatically in recent years, resulting in complex adverse effects on the ecosystem and on human health. These negative impacts on soil and groundwater with resulting effects on quality of life have generated increasing regulatory and economic requirements to monitor and treat pollution in the environment with reliable and cost-effective monitoring solutions for soil and sediments: for example, the Integrated Prevention and Pollution Control (IPPC) Directive, 1996; the Landfill Directive, 1999; and the Water Framework Directive, 2000. New, effective low-cost tools are needed for monitoring pollution and detecting trends over time (Panayiotou et al., 2005).

5.1 Examples of deployments

In the scientific literature, there are not many examples of pervasive networks of sensors to monitor contaminants in soil as this kind of assessment is generally conducted as part of traditional measurement campaigns. Pervasive networks of sensors are generally used to monitor soil moisture as part of smart agriculture applications (Aqeel ur et al., 2014). Wark et al. (2007) adopts a network of custom complementary metal oxide semiconductor (CMOS) camera modules and ECH2O capacitance-based sensors that measure the surrounding soil's volumetric water content as part of a pilot scheme in Australia. Hasan et al. (2013) and Hasan et al. (2015) adopted an RFID-based passive sensor for pervasive soil moisture monitoring: permittivity of soil changes with variations in moisture content, causing a controlled change in the signal response of an RFID tag. Kim et al. (2014) use a similar technology but this time applying it to an inkjet-printed passive soil moisture sensor on paper. This system has the potential to be implemented to also monitor nutrient or salt parameters leached out from the soil and to measure soil water content (Bogena et al., 2010).

As regards the pervasive monitoring of contaminants in soil, scientific literature presents examples related to the monitoring of wastes from chemical industries. Continued presence of chemical waste in large quantities in dumping sites can cause environmental hazards such as soil erosion, water pollution, and air pollution. Kumar et al. (2011) propose a monitoring system using a Zigbee WSN to monitor several dump sites. When chemical waste dumped in a particular site reaches a certain threshold, the system raises an alarm to indicate the need to change the dump site. Kumar et al. also suggest a classification of chemical sensors for these purposes in four groups, depending on the principal physics and operating mechanism of the sensor: (i) spectrometry and chromatography, (ii) electrochemical sensors, (iii) mass sensors, and (iv) optical sensors.

Abdullah et al. (2016) and Zia et al. (2013) use a pH or oxidation-reduction potential (ORP) adapter to measure the variation of pH in soil due to the variation in the concentrations of salts and nutrients because of contaminants or agricultural activities. WSNs have been used to measure the nitrate propagation through soils in parts of the USA: although sensor nodes have become inexpensive, the authors note that the high cost of some sensors (e.g., ammonium) is a major barrier for dense deployment of WSNs (Ramanathan et al., 2006).

The WSN technology was also used by Lin et al. (2007) to monitor soil contamination as a result of the application of commercial N, P, and K fertilizers in agriculture. The WSN technology was used to measure soil nutrients to control the use of fertilizers in order to maximize crop productivity while containing the contamination of soil with resulting negative impacts on surface and groundwater. This approach utilized electrochemical sensors using an ion-selective electrode (ISE), with glass or a polymer membrane, or an ion-selective field effect transistor (ISFET). Ion-selective membranes are able to detect most of the important soil nutrients, including NO_3, K, Na, Ca, Mg, and Cl. In general, laboratory tests of ISE or ISFET sensors have shown good performances and a strong linear relationship with standard methods (Lin et al., 2007).

5.2 Challenges, opportunities, and recommendations

The main challenge identified in this section is actually the limited number of applications of WSNs to monitor contaminants in soil. The application of WSNs to monitor soils is mostly related to smart agriculture practices rather than to perform a pervasive assessment of specific environmental contaminants. As such, applications to pervasively monitor soil humidity and soil nutrients, both with the purpose of increasing crop production, are mostly present in scientific literature. A challenge that has been identified in literature and that may also be relevant to the other fields is the communication range of WSNs. This might have important implications for the long-distance deployment of WSNs in developing countries, which might not have well developed Wi-Fi/cable network infrastructures. Many practical deployments related to farming or water quality monitoring in the developing countries may require sensor monitoring over long distances (Bagula et al., 2012).

6 Conclusions

WSNs have gained tremendous attention in the past decade as an option to monitor the diversity and quantity of chemicals released into the environment due to progress in miniaturization and technological advancements in recent years. Before designing and installing any WSN system, it is necessary to understand its physical environment and deployment in detail, so domain knowledge is of paramount importance. There are, though, some common key challenges and issues that the reader needs to be aware of in order to design an effective pervasive deployment of environmental sensors (Aqeel ur et al., 2014):

- *Energy consumption*: Each node of a WSN requires energy for data processing and transmission, event detection, and potentially data routing (Willig, 2006; Kumar et al., 2015). As such, a WSN requires a proper energy-management strategy in hardware and software in order to optimize the size and duration of a limited and finite energy source, e.g., alkaline batteries or lithium cells, if not powered from mains. The literature also shows use cases where it was possible to avoid this issue by using renewable energy sources such as solar power or kinetic energy (Aqeel ur et al., 2014).
- *Data acquisition, sampling, and transmission*: Data acquisition, sampling, and transmission require power. With limited energy resources, it is critical to optimize data collection and sampling rate in order to collect the desired data without wasting energy, as frequent data acquisition will cause a huge amount of packets being transmitted and this exhausts the energy of the batteries quickly (Alippi et al., 2009). Various wireless transmission technologies and strategies for conserving the energy are available, including local storage of data at sensor nodes, intelligent transmission of data, and sleep mode.
- *Fault tolerance*: Sensor nodes placed in an open harsh environment are prone to physical damage, blockage, and interference. This potential liability could lead to the failure of a sensor and affect the overall work of the WSNs. Generally, different strategies are available to make WSNs more resilient to this kind of failures, such as the use of redundant sensor nodes, reorganization of sensor network, and overlapped sensing regions.
- *Sensor node housing*: The purpose of the environmental enclosure is to make the sensor node resistant to harsh environmental factors such as heat and rain and physical mishandling by humans or animals. The choice of the environmental enclosure, though, is of critical importance as it could have an impact on the performances of the sensors.
- *Sensor placement*: The placement of sensors in a WSN is an important issue regarding the design of the network, algorithms, topologies used, and the parameters being sensed. It needs to be carefully designed and smartly implemented so that a WSN that may work reliably and autonomously could be established. Sensors should be located in positions to collectively cover the area of interest and placed in position and altitudes so as to measure the parameters without obstructions (Aqeel ur et al., 2014).

References

Abdullah, A., Enazi, S.A., Damaj, I., 2016. AgriSys: A smart and ubiquitous controlled-environment agriculture system. In: 2016 3rd MEC International Conference on Big Data and Smart City (ICBDSC), 15–16 March 2016, pp. 1–6.

Aiello, L.M., Schifanella, R., Quercia, D., Aletta, F., 2016. Chatty maps: constructing sound maps of urban areas from social media data. R. Soc. Open Sci. 3, 150690.

Alippi, C., Anastasi, G., Di Francesco, M., Roveri, M., 2009. Energy management in wireless sensor networks with energy-hungry sensors. IEEE Instrum. Meas. Mag. 12, 16–23.

Aqeel ur, R., Abbasi, A.Z., Islam, N., Shaikh, Z.A., 2014. A review of wireless sensors and networks' applications in agriculture. Comput. Stand. Inter. 36, 263–270.

Atzmueller, M., Becker, M., Doerfel, S., Hotho, A., Kibanov, M., Macek, B.-E., Mitzlaff, F., Mueller, J., Scholz, C., Stumme, G., 2012. Ubicon: observing physical and social activities.

In: IEEE International Conference on Green Computing and Communications. IEEE, pp. 317–324.

Bagula, A., Zennaro, M., Inggs, G., Scott, S., Gascon, D., 2012. Ubiquitous sensor networking for development (USN4D): an application to pollution monitoring. Sensors 12, 391–414.

Becker, M., Caminiti, S., Fiorella, D., Francis, L., Gravino, P., Haklay, M.M., Hotho, A., Loreto, V., Mueller, J., Ricchiuti, F., 2013. Awareness and learning in participatory noise sensing. PLoS One 8, e81638.

Bhardwaj, J., Gupta, K.K., Gupta, R., 2015. A review of emerging trends on water quality measurement sensors. In: 2015 International Conference on Technologies for Sustainable Development (ICTSD), 4–6 February 2015, pp. 1–6.

Bogena, H., Herbst, M., Huisman, J., Rosenbaum, U., Weuthen, A., Vereecken, H., 2010. Potential of wireless sensor networks for measuring soil water content variability. Vadose Zone J. 9, 1002–1013.

Brienza, S., Galli, A., Anastasi, G., Bruschi, P., 2015. A low-cost sensing system for cooperative air quality monitoring in urban areas. Sensors 15, 12242–12259.

Cavaliere, A., Carotenuto, F., Di Gennaro, F., Gioli, B., Gualtieri, G., Martelli, F., Matese, A., Toscano, P., Vagnoli, C., Zaldei, A., 2018. Development of low-cost air quality stations for next generation monitoring networks: calibration and validation of PM2. 5 and PM10 sensors. Sensors 18, 2843.

Chung, W.-Y., Yoo, J.-H., 2015. Remote water quality monitoring in wide area. Sensors Actuators B Chem. 217, 51–57.

Coletta, C., Kitchin, R., 2017. Algorhythmic governance: regulating the 'heartbeat'of a city using the Internet of Things. Big Data Soc. 4 (2).

Cordero, J.M., Borge, R., Narros, A., 2018. Using statistical methods to carry out in field calibrations of low cost air quality sensors. Sensors Actuators B Chem. 267, 245–254.

Cui, Y., Wei, Q., Park, H., Lieber, C.M., 2001. Nanowire nanosensors for highly sensitive and selective detection of biological and chemical species. Science 293, 1289–1292.

D'Hondt, E., Stevens, M., Jacobs, A., 2013. Participatory noise mapping works! An evaluation of participatory sensing as an alternative to standard techniques for environmental monitoring. Pervasive Mob. Comput. 9, 681–694.

European Commission, 2002. Directive 2002/49/EC Relating to the Assessment and Management of Environmental Noise (the Environmental Noise Directive – END).

Farrés, J.C., 2015. Barcelona noise monitoring network. Proc. EuroNoise, 218–220.

Gao, M., Cao, J., Seto, E., 2015. A distributed network of low-cost continuous reading sensors to measure spatiotemporal variations of PM2. 5 in Xi'an, China. Environ. Pollut. 199, 56–65.

Hachem, S., Mallet, V., Ventura, R., Pathak, A., Issarny, V., Raverdy, P., Bhatia, R., 2015. Monitoring noise pollution using the urban civics middleware. In: 2015 IEEE First International Conference on Big Data Computing Service and Applications, 30 March–2 April 2015, pp. 52–61.

Haji Gholizadeh, M., Melesse, A.M., Reddi, L., 2016. Spaceborne and airborne sensors in water quality assessment. Int. J. Remote Sens. 37, 3143–3180.

Han, M., Gao, X., Su, J.Z., Nie, S., 2001. Quantum-dot-tagged microbeads for multiplexed optical coding of biomolecules. Nat. Biotechnol. 19, 631.

Hao, T., Xing, G., Zhou, G., 2013. iSleep: unobtrusive sleep quality monitoring using smartphones. In: Proceedings of the 11th Acm Conference on Embedded Networked Sensor Systems. ACM, p. 4.

Hasan, A., Bhattacharyya, R., Sarma, S., 2013. A monopole-coupled RFID sensor for pervasive soil moisture monitoring. In: 2013 IEEE Antennas and Propagation Society International Symposium (APSURSI), 7–13 July 2013, pp. 2309–2310.

Hasan, A., Bhattacharyya, R., Sarma, S., 2015. Towards pervasive soil moisture sensing using RFID tag antenna-based sensors. In: 2015 IEEE International Conference on RFID Technology and Applications (RFID-TA), 16–18 September 2015, pp. 165–170.

Huang, X., Yi, J., Chen, S., Zhu, X., 2015. A wireless sensor network-based approach with decision support for monitoring lake water quality. Sensors 15.

Huang, P., Jin, Y., Hou, D., Yu, J., Tu, D., Cao, Y., Zhang, G., 2017. Online classification of contaminants based on multi-classification support vector machine using conventional water quality sensors. Sensors 17.

Jeffrey Yang, Y., Haught, R.C., Goodrich, J.A., 2009. Real-time contaminant detection and classification in a drinking water pipe using conventional water quality sensors: techniques and experimental results. J. Environ. Manag. 90, 2494–2506.

Jeon, J., Kim, J.H., Lee, B.C., Kim, S.D., 2008. Development of a new biomonitoring method to detect the abnormal activity of Daphnia magna using automated grid counter device. Sci. Total Environ. 389, 545–556.

Jiao, W., Hagler, G., Williams, R., Sharpe, R., Brown, R., Garver, D., Judge, R., Caudill, M., Rickard, J., Davis, M., 2016. Community air sensor network (CAIRSENSE) project: evaluation of low-cost sensor performance in a suburban environment in the southeastern United States. Atmos. Meas. Tech. 9, 5281–5292.

Kanjo, E., 2010. Noisespy: a real-time mobile phone platform for urban noise monitoring and mapping. Mob. Netw. Appl. 15, 562–574.

Kaufman, A., Williams, R., Barzyk, T., Greenberg, M., O'shea, M., Sheridan, P., Hoang, A., Ash, C., Teitz, A., Mustafa, M., 2017. A citizen science and government collaboration: developing tools to facilitate community air monitoring. Environ. Justice 10, 51–61.

Kim, S., Le, T., Tentzeris, M.M., Harrabi, A., Collado, A., Georgiadis, A., 2014. An RFID-enabled inkjet-printed soil moisture sensor on paper for "smart" agricultural applications. In: Sensors, 2014 IEEE, 2–5 November 2014, pp. 1507–1510.

Kroll, D., 2006. Securing Our Water Supply: Protecting a Vulnerable Resource.

Kruse, P., 2018. Review on water quality sensors. J. Phys. D. Appl. Phys. 51, 203002.

Kumar, K.N., Prabakaran, R., Dhulipala, V.S., Ranjith, P., 2011. Future sensors and utilization of sensors in chemical industries with control of environmental hazards. In: Proceedings of the 2nd International Conference on Environmental Science and Development, pp. 224–228.

Kumar, P., Morawska, L., Martani, C., Biskos, G., Neophytou, M., Di Sabatino, S., Bell, M., Norford, L., Britter, R., 2015. The rise of low-cost sensing for managing air pollution in cities. Environ. Int. 75, 199–205.

Leigh, C., Alsibai, O., Hyndman, R.J., Kandanaarachchi, S., King, O.C., Mcgree, J.M., Neelamraju, C., Strauss, J., Talagala, P.D., Turner, R.D.R., Mengersen, K., Peterson, E. E., 2019. A framework for automated anomaly detection in high frequency water-quality data from in situ sensors. Sci. Total Environ. 664, 885–898.

Lin, J., Wang, M., Zhang, M., Zhang, Y., Chen, L., 2007. Electrochemical sensors for soil nutrient detection: opportunity and challenge. In: International Conference on Computer and Computing Technologies in Agriculture. Springer, pp. 1349–1353.

Liu, S., Che, H., Smith, K., Chang, T., 2015. A real time method of contaminant classification using conventional water quality sensors. J. Environ. Manag. 154, 13–21.

Lu, H., Pan, W., Lane, N.D., Choudhury, T., Campbell, A.T., 2009. SoundSense: scalable sound sensing for people-centric applications on mobile phones. In: Proceedings of the 7th International Conference on Mobile Systems, Applications, and Services. ACM, Poland. Kraków.

Lu, H., Frauendorfer, D., Rabbi, M., Mast, M.S., Chittaranjan, G.T., Campbell, A.T., Gatica-Perez, D., Choudhury, T., 2012. Stresssense: detecting stress in unconstrained acoustic environments using smartphones. In: Proceedings of the 2012 ACM Conference on Ubiquitous Computing. ACM, pp. 351–360.

Maisonneuve, N., Stevens, M., Niessen, M.E., Hanappe, P., Steels, L., 2009a. Citizen noise pollution monitoring. In: Proceedings of the 10th Annual International Conference on Digital Government Research: Social Networks: Making Connections between Citizens, Data and Government. Digital Government Society of North America, Puebla, Mexico.

Maisonneuve, N., Stevens, M., Niessen, M.E., Steels, L., 2009b. NoiseTube: Measuring and Mapping Noise Pollution with Mobile Phones. Information Technologies in Environmental Engineering. Springer.

Malings, C., Tanzer, R., Hauryliuk, A., Kumar, S.P.N., Zimmerman, N., Kara, L.B., Presto, A. A., Subramanian, R., 2019. Development of a general calibration model and long-term performance evaluation of low-cost sensors for air pollutant gas monitoring. Atmos. Meas. Tech. 12, 903–920.

Martinez, K., Hart, J.K., Ong, R., 2004. Environmental sensor networks. Computer 37, 50–56.

Morawska, L., Thai, P.K., Liu, X., Asumadu-Sakyi, A., Ayoko, G., Bartonova, A., Bedini, A., Chai, F., Christensen, B., Dunbabin, M., Gao, J., Hagler, G.S.W., Jayaratne, R., Kumar, P., Lau, A.K.H., Louie, P.K.K., Mazaheri, M., Ning, Z., Motta, N., Mullins, B., Rahman, M. M., Ristovski, Z., Shafiei, M., Tjondronegoro, D., Westerdahl, D., Williams, R., 2018. Applications of low-cost sensing technologies for air quality monitoring and exposure assessment: how far have they gone? Environ. Int. 116, 286–299.

Navarro, J.M., Tomas-Gabarron, J.B., Escolano, J., 2017. A big data framework for urban noise analysis and management in smart cities. Acta Acust. United Ac. 103, 552–560.

Panayiotou, C.G., Fatta, D., Michaelides, M.P., 2005. Environmental Monitoring Using Wireless Sensor Networks. University of Cyorus, Dept of Electrical Engineering.

Qin, Y., Alam, A.U., Pan, S., Howlader, M.M.R., Ghosh, R., Hu, N.-X., Jin, H., Dong, S., Chen, C.-H., Deen, M.J., 2018. Integrated water quality monitoring system with pH, free chlorine, and temperature sensors. Sensors Actuators B Chem. 255, 781–790.

Rai, A.C., Kumar, P., Pilla, F., Skouloudis, A.N., Di Sabatino, S., Ratti, C., Yasar, A., Rickerby, D., 2017. End-user perspective of low-cost sensors for outdoor air pollution monitoring. Sci. Total Environ. 607-608, 691–705.

Ramanathan, N., Balzano, L., Estrin, D., Hansen, M., Harmon, T., Jay, J., Kaiser, W., Sukhatme, G., 2006. Designing wireless sensor networks as a shared resource for sustainable development. In: 2006 International Conference on Information and Communication Technologies and Development. IEEE, pp. 256–265.

Rana, R.K., Chou, C.T., Kanhere, S.S., Bulusu, N., Hu, W., 2010. Ear-phone: an end-to-end participatory urban noise mapping system. In: Proceedings of the 9th ACM/IEEE International Conference on Information Processing in Sensor Networks. ACM, pp. 105–116.

Reece, S., Kaufman, A., Hagler, G., Williams, R., 2017. Low-Cost Sensor POD Design Considerations. EM. (accessed on 24 September 2018), pp. 24–30.

Rossi, M., Feese, S., Amft, O., Braune, N., Martis, S., Tröster, G., 2013. AmbientSense: A real-time ambient sound recognition system for smartphones. In: 2013 IEEE International Conference on Pervasive Computing and Communications Workshops (PERCOM Workshops), 18–22 March 2013, pp. 230–235.

Santini, S., Ostermaier, B., Vitaletti, A., 2008. First experiences using wireless sensor networks for noise pollution monitoring. In: Proceedings of the Workshop on Real-World Wireless Sensor Networks. ACM, pp. 61–65.

Segura-Garcia, J., Felici-Castell, S., Perez-Solano, J.J., Cobos, M., Navarro, J.M., 2015. Low-cost alternatives for urban noise nuisance monitoring using wireless sensor networks. IEEE Sensors J. 15, 836–844.

Stevens, M., D'Hondt, E., 2010. Crowdsourcing of pollution data using smartphones. Workshop on Ubiquitous Crowdsourcing. pp. 1–4.

Sun, L., Wong, K., Wei, P., Ye, S., Huang, H., Yang, F., Westerdahl, D., Louie, P., Luk, C., Ning, Z., 2016. Development and application of a next generation air sensor network for the Hong Kong marathon 2015 air quality monitoring. Sensors 16, 211.

US EPA, 2003. Planning for and Responding to Drinking Water Contamination Threats and Incidents. EPA 817-D-04-001, Washington, DC.

US EPA, 2005. Water sentinel system architecture. In: EPA, U. (Ed.), Office of Water, Office of Ground Water and Drinking Water. EPA 817-D-05-003, Washington, DC.

US EPA, 2008. Water security initiative Cincinnati Pilot Post implementation system status: covering the pilot period: December 2005 through December 2007. In: EPA, U. (Ed.), Office of Water, Office of Ground Water and Drinking Water. EPA 817-R-08-004, Washington, DC.

Vikesland, P.J., 2018. Nanosensors for water quality monitoring. Nat. Nanotechnol. 13, 651–660.

Wark, T., Corke, P., Sikka, P., Klingbeil, L., Guo, Y., Crossman, C., Valencia, P., Swain, D., Bishop-Hurley, G., 2007. Transforming agriculture through pervasive wireless sensor networks. IEEE Pervasive Comput. 6, 50–57.

White, J., Thompson, C., Turner, H., Dougherty, B., Schmidt, D.C., 2011. WreckWatch: automatic traffic accident detection and notification with smartphones. Mob. Netw. Appl. 16, 285–303.

White, R., Paprotny, I., Doering, F., Cascio, W., Solomon, P., Gundel, L., 2012. Sensors and 'apps' for community-based atmospheric monitoring. EM: Air and Waste Management Association's Magazine for Environmental Managers 5, 36–40.

Williams, R., Kilaru, V., Snyder, E., Kaufman, A., Dye, T., Rutter, A., Russell, A., Hafner, H., 2014. Air Sensor Guidebook. US Environmental Protection Agency.

Willig, A., 2006. Wireless sensor networks: concept, challenges and approaches. Elektrotech. Informationstech. 123, 224–231.

Zia, H., Harris, N.R., Merrett, G.V., Rivers, M., Coles, N., 2013. The impact of agricultural activities on water quality: a case for collaborative catchment-scale management using integrated wireless sensor networks. Comput. Electron. Agric. 96, 126–138.

Zigel, Y., Litvak, D., Gannot*, I., 2009. A method for automatic fall detection of elderly people using floor vibrations and sound—proof of concept on human mimicking doll falls. IEEE Trans. Biomed. Eng. 56, 2858–2867.

Zimmerman, N., Presto, A.A., Kumar, S.P.N., Gu, J., Hauryliuk, A., Robinson, E.S., Robinson, A.L., Subramanian, R., 2018. A machine learning calibration model using random forests to improve sensor performance for lower-cost air quality monitoring. Atmos. Meas. Tech. 11, 291–313.

Zwicker, E., Fastl, H., 2013. Psychoacoustics: Facts and Models. Springer Science & Business Media.

Challenges and current approaches toward environmental monitoring of nanomaterials

4

Geert Cornelis[a], Jani Tuoriniemi[a], Manuel Montaño[b], Stephan Wagner[c], Julian A. Gallego-Urrea[d], Karin Mattsson[d], and Andreas Gondikas[e]
[a]Department of Soil and Environment, Swedish University of Agricultural Sciences, Uppsala, Sweden, [b]Huxley College of the Environment, Western Washington University, Bellingham, WA, United States, [c]Department of Analytical Chemistry, Helmholtz-Centre for Environmental Research—UFZ, Leipzig, Germany, [d]Department of Marine Sciences, University of Gothenburg, Fiskebäckskil, Sweden, [e]Faculty of Geology and Geo-environment, National and Kapodistrian University of Athens, Athens, Greece

1 Introduction

Engineered nanomaterials (ENMs) have found their way into our daily lives and also into our environment. This fact has sparked a decade of research into the potentially harmful effects of ENMs and also into their environmental pathways. These investigations indeed revealed that ENMs may in some cases pose harmful effects to organisms, but the concentrations at which these effects occur vary widely, not only with the properties of the ENMs but also with the nature of the specific environmental compartment where the ENM was emitted into (Klaine et al., 2008).

Analytical tools to measure the concentration of ENMs in complex matrices have been developed parallel to hazard and exposure investigations. It became quite clear that advanced tools were required to confirm ENM doses, even in standard tests where it can be argued that the matrix is simple, relative to real-world samples (Petersen et al., 2014; Handy et al., 2012). ENMs impose more stringent requirements in terms of dose confirmation compared to conventional chemicals, because where a mere concentration determination suffices for the latter, the used ENM material needs to be characterized in terms of size dimensions, composition crystal structure, etc., to establish exactly which material the organisms have been exposed to (Petersen et al., 2014). In addition, ENMs may homoaggregate, heteroaggregate, dissolve, attach to surfaces, acquire or lose coating, or even change core composition in both aquatic and terrestrial tests, and these changes in many cases affect the ENM hazard (Petersen et al., 2014; Velzeboer et al., 2008; Handy and Shaw, 2007; Bone et al., 2012). Analytical tools, therefore, need to be able to capture those dynamic changes during the test and not only at the start, at least to aid interpreting the endpoint measurements and potentially also to calculate more accurate dose-response curves (Petersen et al., 2014; Velzeboer et al., 2008; Handy and Shaw, 2007).

Monitoring Environmental Contaminants. https://doi.org/10.1016/B978-0-444-64335-3.00003-7

Environmental monitoring of ENM comprises ENM measurements in the most complicated field samples. Risk assessment of ENMs has hitherto relied heavily on multimedia mass flow modeling (e.g. (Gottschalk et al., 2013; Sun et al., 2016; Sun et al., 2014; Gottschalk and Nowack, 2011; Gottschalk et al., 2009; Arvidsson et al., 2012; Meesters et al., 2014; Meesters et al., 2016; Praetorius et al., 2012)). These have estimated predicted environmental concentrations (PECs) in different environmental compartments, based on known or assumed production volumes and generalized information on their transport within these compartments. Actual ENM measurements in the field have, however, not been produced to confirm these estimates or provide effective concentrations that can be compared with hazard levels for risk calculations. There are conceptual explanations but certainly also technical ones (Nowack, 2017; Nowack et al., 2015; Von der Kammer et al., 2012). While ENM dose confirmation in standard tests is already challenging, the analysis of ENMs in real-world samples imposes even more demands that go beyond analytical complications such as sample preparation that also complicate measurements of other contaminants (Von der Kammer et al., 2012).

There are two distinct goals of environmental ENM monitoring, each having specific technical challenges:

1. Quantifying the total concentration of ENMs in a given field sample, which requires specificity and low concentration detection limits.
2. Analyzing the relative proportion of different forms of occurrence of ENMs, which requires versatility and low size detection limits.

Total concentration analysis seeks to provide, in the first instance, mass concentrations of ENMs in a given natural water, soil, sediment, etc. ENMs need to be distinguished from other materials in the sample such as dissolved ions or naturally occurring particles, but the forms in which these intact ENMs occur is of no immediate concern. This approach likely serves risk assessments well as it provides an environmental concentration that can be compared to predicted no-effect concentrations that are to date never expressed in terms of the specific forms of occurrence of ENMs. However, it is clear that a material-specific and sensitive approach is required to detect the projected low ENM concentrations in environmental media against a background of natural particles. The main tool to detect total ENM concentration during the initial stages of research into environmental effects of ENMs has, therefore, been inductively coupled plasma-mass spectrometry (ICPMS). This is one of the most sensitive techniques available that can detect specific materials in complex samples, though only for relatively heavy, inorganic elements. This is one explanation why much of the ENM studies were initially done on ENMs consisting of relatively heavy elements such as Ag, Au, or CeO_2 (see e.g., (Klaine et al., 2008)). However, solely using ICPMS does not distinguish elements occurring as ENMs, dissolved ions, or bulk materials. Should regulatory limit values ever be expressed for ENMs, legally binding ENM detection would always require some form of size analysis, given that the legal definition of ENMs will mostly be based on their size (European Commission, 2011). Using ICPMS in the single-particle mode (spICPMS), as discussed further, has in part solved this complication (See also Fig. 1). It combines the specificity and sensitivity of

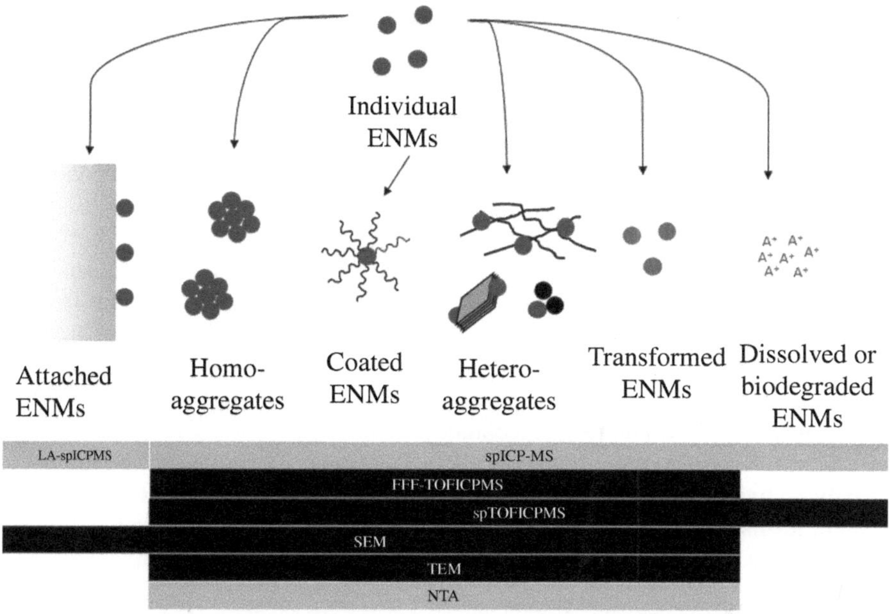

Fig. 1 The (conceptual) different forms of occurrence of ENMs in environmental samples and how they can be distinguished from individual ENMs by some of the techniques are discussed in this chapter. *Black and gray shades* mean that the technique can respectively cannot distinguish the different forms from individual ENMs. White shades mean the form cannot be detected.

ICPMS with a mass-based estimate of their size, also distinguishing ENMs from dissolved ions. spICPMS is conversely less well suited for detecting the multiple possible forms of occurrence of ENMs and is not suitable for detecting carbon-based ENMs.

Detecting the different forms of ENM occurrence is a more ambitious goal than a mere total concentration determination. ENMs can attach to pore walls or other large environmental surfaces, heteroaggregate with ubiquitous naturally occurring particles, homoaggregate with themselves, be coated with dissolved organic matter (DOM), or be transformed into ENMs with a different composition (Cornelis et al., 2014) (Fig. 1). In addition, highly reactive conditions may prevail that change the ENM material composition entirely. A classic example is how low redox potentials and high sulfide concentrations in wastewater treatment plants rapidly transform metallic silver particles into silver sulphide particles (Kaegi et al., 2011; Doolette et al., 2013). All these forms have a different effect on organisms, but the current risk assessment tools are insufficiently developed to take them into account. A detailed analysis of ENM forms, therefore, currently occurs mostly during in-depth mechanistic studies, where the ENMs are added to environmental media during laboratory studies. The ENM concentration can thus often be set to levels above the detection limits of the analytical tool used, because the focus is on the diverse forms of ENM occurrence. The concentration detection limit is therefore not a major concern. Major tools to study the forms of occurrence of ENMs have therefore been electron microscopy-based techniques, as discussed further. These could be of the size of the smallest ENMs and at the same time detect the different forms of occurrence of ENMs (Fig. 1). Microscopic tools are conversely less well suited to detect environmentally realistic ENM concentrations.

It thus appears that no single technique exists that can both provide detailed information on the form of occurrence and do this at environmentally realistic concentrations, although it will be discussed how specific developments, e.g., multielement single particle time-of-flight ICPMS (spICPTOFMS) are increasingly achieving this. The goal of this chapter is not to provide an exhaustive overview of all the analytical tools that exist to characterize ENMs in ecotoxicological or environmental samples. The reader is referred to excellent reviews on that topic (Von der Kammer et al., 2012; Domingos et al., 2009; Farré et al., 2011). Instead, only those tools are discussed that can provide the currently best possible answer to specific questions in environmental ENM monitoring. This entails that techniques typically used to characterize homogeneous ENM dispersions (e.g., dynamic light scattering, DLS) will not be discussed, but only techniques able to analyze environmentally relevant concentrations and/or different forms of occurrence will be considered. It will be discussed how these techniques address the following aspects that are important for environmental monitoring:

- Specificity to distinguish ENMs from naturally occurring particles
- Concentration detection limits
- Size detection limits
- The ability to measure number concentrations
- Minimizing sample pretreatment

Material specificity allows detecting ENM in the presence of backgrounds that are often ubiquitous in environmental samples. Environmental monitoring of carbon-based NMs has been developed much less in this respect compared to inorganic ENMs. In part, this is because production volumes of, e.g., carbon nanotubes and fullerenes were much less compared to inorganic ENMs such as TiO_2 (Piccinno et al., 2012), but mostly because detection of C-based ENMs could not rely on ICPMS based approaches, but on other detection methodologies such as infrared, which are significantly less sensitive (Petersen et al., 2016). Moreover, molecular detection of ENMs among a multitude of organic compounds in environmental samples imposes even higher requirements compared to inorganic ENMs. However, recent enhanced focus on much more ubiquitous C-based ENM materials such as graphene (Arvidsson et al., 2018) or nanoplastics (Horton et al., 2017) also enhances the need for detecting C-based ENM materials.

Low concentration limits are important, as discussed, to monitor environmentally relevant, low ENM concentrations. It may also be important to use low concentrations during laboratory effects of fate tests, because increasing ENM concentration promotes processes such as homoaggregation, which compromises the realism of the environmental systems simulated in the laboratory (Petersen et al., 2014).

Size detection limits vary tremendously with the material the ENM is made from. ENM made out of heavy elements such as Au ENM often have the best detection limits for most techniques because they have a low ionization potential, which enhances ICP-based detection (Lee et al., 2014); good optical properties, which favors scattering detection (Kaszuba et al., 2008); and are electron-rich, which favors resolution in electron microscopy applications (Williams and Carter, 1996). In the case where size is calculated from the mass of an element per particle, as in spICPMS, the size detection limit also determines the limit below which ENMs can no longer be distinguished from dissolved ions (Lee et al., 2014).

The ability to measure number concentrations is often mentioned as a requirement for ENMs based on the EC definition of ENMs that is expressed in terms of a certain percentage of the particle size distribution that needs to be below the arbitrary size limit of 100 nm (European Commission, 2011). While number concentrations are thus legally important to judge whether particular materials are subject to nanospecific regulation, it can be deemed, at least currently, less important for environmental monitoring, because predicted no-effect concentrations are as a rule expressed in terms of mass concentrations. Moreover, there is a direct relation between number and mass using the average size in the case of monodisperse ENM suspensions. Some studies argue that number concentration is a better dose metric to express hazard to ENM, but this is usually in an inhalation toxicity context (e.g., Schmid and Stoeger, 2016; Donaldson et al., 2013). Early aquatic ecotoxic studies suggested the same (Van Hoecke et al., 2008, 2011), but there actually is still not enough evidence to support number concentration as the best dose metric for expressing ENM hazard (Skjolding et al., 2016). There are better arguments to express exposure predictions based on number concentration because most kinetic processes controlling their concentrations are proportional to number concentrations, not mass (Petosa et al., 2010). This is mostly important for polydisperse ENMs where there is no straightforward relationship between mass and number concentration. However, it is not common practice, neither in hazard assessments nor fate predictions, to work with number concentrations. This is because there is no conservation law for numbers as there is for mass, but likely also because routine analytical tools are unable to analyze number concentrations and the techniques that can measure number concentrations, such as spICPMS, do so with varying precision (Liu et al., 2017; Laborda et al., 2013a; Tuoriniemi et al., 2014a; Gallego-Urrea et al., 2010). It is therefore important to explore our current capability to provide number concentrations.

Minimizing sample pretreatment reduces the risk of measurement artifacts and enhances throughput thus reducing costs and in some cases even reduces detection limits. Each of the environmental compartments in which ENMs may be emitted poses different challenges in terms of sample treatment that need to be overcome to minimize artifacts during environmental monitoring. Some techniques discussed in this chapter can provide a wealth of information on ENMs, but only following a costly sample pretreatment. It is unlikely that routine environmental monitoring will therefore adopt these techniques despite their merits. The chapter is therefore oriented around different environmental compartments, and it will be discussed how analytical development in some cases has managed to reduce sample pretreatment and thus lower costs and enhance throughput.

The current chapter is a snapshot in time. Development is ongoing, because the requirements of ENM analysis are expected to only increase. A current development is that ENMs are becoming increasingly complex consisting, for instance, of hybrid materials. Some of the remaining challenges therefore needs to be combined with foreseen challenges.

2 Analytical techniques for environmental monitoring of ENMs

2.1 Scanning and transmission electron microscopy

Scanning electron microscopy (SEM) and transmission electron microscopy (TEM) are two of the most widely used characterization techniques in nanotechnology because of the possibility to characterize individual particles in detail. At the same time, their usefulness for routine monitoring of trace particle contamination is limited because of the small volumes of sample that can be probed within a reasonable time.

During SEM analysis, an electron beam is focused and deflected to scan an area of the sample in a raster fashion. Detectors are located above the sample in order to measure the secondary electron current (SE) and backscattered electron current (BSE). Qualitative analysis of elemental composition can be carried out with energy dispersive spectroscopy (EDS) or wavelength dispersive spectroscopy (WDS). The contrast in SE imaging is mainly topographical, while BSE mainly gives atomic number contrast where heavy metal containing particles can be visualized as bright spots against the dark matrix. The depth of focus in SEM is high, and the height of the focal plane can be adjusted within a range of centimeters, which allows imaging of several centimeter-sized objects with little prior sample preparation (Mavrocordatos et al., 2007).

High vacuum and sample conductivity are required in normal SEM where the electron gun would otherwise ionize air, which would in turn conduct away the electron charge thus impeding imaging because charge would accumulate on insulating samples. Nonconducive environmental samples thus need to be sputtered with a conducive material, most often Au or Pt (Mavrocordatos et al., 2007). The necessary drying step is known to introduce artifacts in SEM analysis of particles (Mavrocordatos et al., 2007; Bootz et al., 2004). In variable-pressure SEM (VPSEM), it is possible to keep the atmospheric pressure within the sample chamber higher than what is required during normal SEM, and the sample drying thus becomes less crucial because there is less need for sputtering nonconductive samples.

The main advantage of SEM is that a specimen area may be scanned within a reasonable amount of time, yet the resolution is often not sufficient for the analysis of nanoparticles. The resolution can in SEM imaging reach 1–2 nm for good conductors under high vacuum (Williams and Carter, 1996). However, for e.g., layers of soils or dusts imaged in the VPSEM mode, the smallest objects that are practical to examine with SE are typically 10 -- 20 nm. The resolution in BSE is usually lower than with SE depending on the penetration depth of the electron beam into the sample. The possibilities of using BSE imaging as a tool for measuring number concentrations of engineered nanoparticles in soils and related samples are discussed further in this chapter.

TEM analysis is analogous to optical microscopy, in that an image is formed by the transmitted signal, but an electron beam is used to illuminate samples during TEM. The resolution can reach below 1 nm (Williams and Carter, 1996; Mavrocordatos et al., 2007). The crystal phase and habit of individual particles can be determined with electron diffraction (ED) and high resolution (HR) TEM; and similarly, SEM and EDS can be used to determine particle composition.

However, the samples are required to be thin, usually below 100 nm in thickness, and the high resolution implies that very small areas (typically less than 1 μm^2) of a sample can be analyzed at a time (Williams and Carter, 1996; Mavrocordatos et al., 2007). The TEM instrument is often operated manually; thus examining the whole area of a sample of unknown composition is unrealistically time-consuming. The method may therefore only be used to study ENMs that have been isolated and concentrated beforehand. Although several approaches have been reported, there is no protocol for sample pretreatment that can be applied universally prior to TEM or SEM, and the treatment is adjusted to the characteristics of the aqueous sample.

The size detection limit of electron microscopy techniques is highly dependent on the sample properties and operational conditions. For example, if an ENM has been identified on the BSE detector and compositional information is acquired with EDS, the latter will vary depending on the acceleration voltage of the beam. Higher voltage results in higher beam penetration, which results in a higher proportion of the underlying material in the acquired compositional information.

Because the samples are required to be thin, usually below 100 nm in thickness, EDS analysis is quantitative in the absence of secondary electrons or X-ray fluorescence. The stringent requirements for the sample mean that a perturbing sample preparation procedure must be carried out, often involving resin embedding and ion beam milling or adsorption of colloids onto a holey carbon film- coated grid (Von der Kammer et al., 2012; Mavrocordatos et al., 2007; Plathe et al., 2010, 2013). Quantitative relations between the concentrations in samples and the numbers of particles counted in images are therefore difficult to establish. The outcome of a search for trace particles is always uncertain as only minuscule volumes of the sample can be examined. However, although TEM is not a technique suited for routine contaminant monitoring, it plays an important role in environmental nanoscience because the mineral phases of trace particles can hardly be determined by any other method.

2.2 Field- flow fractionation

Field- flow fractionation (FFF) is a particle size fractionation technique based on hydrodynamic principles (Giddings et al., 1977; Wahlund, 2013) that has been applied to separate proteins, polymers (Giddings, 1993), natural colloids (von der Kammer et al., 2005), and ENMs (Hagendorfer et al., 2011). Suspended particles are separated by an external field force while flowing through a channel or a hollow fiber membrane. There are a number of different external field forces applied, for instance, a thermal gradient, gravitational force, or a water flow. Most commonly, asymmetric flow field- flow fractionation (AF4) (von der Kammer et al., 2005; Hassellöv et al., 1999) or centrifugal field- flow fractionation (CF3) (Beckett et al., 1988) is applied to particle suspensions. FFF is a sample preparation technique to produce particle size fractions, which can be further analyzed. FFF systems can be hyphenated to various detectors such as DLS and static light scattering, UV/vis, and ICPMS each providing different types of information, e.g., particle size and elemental mass concentration, making FFF techniques a powerful tool for comprehensive characterization of particle suspensions. For more details on the separation principle, we refer to earlier publications (e.g., Baalousha et al., 2006).

Although FFF techniques have a high potential for comprehensive characterization of ENMs in environmental samples, there are also challenges that may limit the routine application of this technique. Under ideal conditions, the separation principle of AF4 and CF3 is solely based on hydrodynamics, in practice, however, particle-particle as well as particle-membrane (only for AF4) interactions may affect the outcome of the separation. These interactions can be controlled by tedious method development (Loeschner et al., 2013; Gigault et al., 2014). To establish FFF methodologies and to introduce quality control parameters, an ISO guideline on FFF method development has been published in 2018 (International Organization for Standardization, 2018). Besides the tedious method development for particle separation, FFF techniques are applicable for separation of stable particle suspensions only. Therefore, in most cases environmental samples cannot be directly fractionated by FFF without a pretreatment like preconcentration, filtration, or extraction of the colloidal fraction because sample components may affect the particle size fractionation (Wagner et al., 2015). The presence of large particles in the suspension could lead to a misinterpretation of results because they pre-elute. Matrix components may mediate interactions between particles and the membrane causing a loss of particles at the membrane surface (Meisterjahn et al., 2016). Particle elution behavior is also shape dependent. For instance, plate-like particles are prone to accumulate and attach at the membrane surface in AF^4. Although particle sizes between 1 nm and 50 μm can theoretically be separated, this is not possible under constant separation conditions. Therefore, separation of poly-disperse particle suspension is limited. AF4 as a stand-alone technique is able to separate particle sizes, but it does not differentiate between particle constituents. An identification of ENMs in environmental matrices is therefore only possible in combination with elemental detection techniques such as ICPMS.

2.3 *IR/Raman microscopy*

In Fourier transform infrared spectroscopy (FT-IR), IR radiation passes through a sample where some of the radiation is absorbed by the samples, and some of it is transmitted. The obtained spectrum represents the molecular absorption and transmission, a fingerprint of the sample that is identified through comparing it with a library. From the spectra, it is also possible to identify aging of the particles by observing the surface oxidation via characteristic bands. There are two operation modes: total reflection, which is mainly used for larger particles (larger than 50 μm) and focal plane array detector based micro FT-IR, which images particles on filters with a resolution down to 10 μm (Simon et al., 2018; Kappler et al., 2016; Löder et al., 2015). It is, however, possible to identify nanosized particles in the form of aggregates or based on a bulk analysis of many particles such as powders (Hernandez et al., 2017). A combination of atomic force microscopy (AFM) and FT-IR, called nano-FT-IR, can give a spectral resolution of 50 nm (Dazzi et al., 2012; Dazzi and Prater, 2017; Meyns et al., 2018).

Raman spectroscopy is a spectroscopic technique based on inelastic scattering from monochromatic light such as laser light. Monochromatic light interacts with the samples, and the photons are absorbed by the sample and re-emitted with a shifted frequency. The shift provides information about vibrational, rotational, and low frequency transitions in molecules. Since Raman microscopes are confocal, a higher resolution than FT-IR can be obtained, and environmental particles down to 1 μm have been identified (Kappler et al., 2016; Imhof et al., 2016). By combining AFM with Raman, a spatial resolution of 10 nm can be achieved (Schmid et al., 2013). Another option is to combine SEM with an insitu Raman, which has a resolution of 200–300 nm (WITec GmbH, 2018) (Fig. 3).

Both Raman and FT-IR are suitable for identification of polymers; however, the samples need to be clean, and it is time consuming. The identification process, i.e., comparing the obtained spectra with a library is fast, but for environmental- weathered particles, the process is more complex since the particle surface may have changed. The Raman signal is less sensitive to surface degradation compared to FT-IR and has a depth resolution of around 1 mm. FT-IR is sensitive to signals from water molecules, so the samples need to be dry, whereas with Raman it is possible to analyze solid, liquid, and gaseous samples. However, with Raman, samples exhibiting fluorescence cannot be measured as these generate noise in the spectra.

2.4 Pyrolysis-GC-MS

Pyrolysis-GC–MS is, since many years, a common technique to measure the concentration of anthropogenic compounds in solid samples such as soils (De Leeuw et al., 1986). Reasonably volatile polymers evaporate, and their decomposition products can be analyzed using hyphenated gas chromatography coupled to a mass spectrometer (GCMS) to distinguish them from other thermal decomposition products. The technique has recently been applied for analyzing the concentration of micro and nanoplastics in solid samples such as soils, organisms (Fischer and Scholz-Böttcher, 2017), and sediments (Fabbri et al., 2000; Dekiff et al., 2014; Fries et al., 2013); the analysis of which otherwise would require extensive sample pretreatment procedures (Bläsing and Amelung, 2018). Aqueous samples can also be analyzed using this technique but need to be filtered first (Fabbri et al., 2000; Dümichen et al., 2017). This approach can replace the time-costly manual identification of plastic particles on a particle per particle basis using optical or IR-based microscopes (Bläsing and Amelung, 2018). The latter technique can, however, also provide size information, whereas Py-GC-MS, without sample pretreatment, cannot distinguish nano, micro or macro plastics. It is, however, possible to analyze associated contaminants by gradually increasing the pyrolysis temperature (Fries et al., 2013).

2.5 Single particle ICPMS

Detecting ENMs in the environment requires methods that possess the requisite sensitivity to monitor at low expected release concentration (ng L^{-1}), and methods that are selective toward specific ENM properties to counteract the ubiquity of naturally occurring nanomaterials. ICPMS is a routine tool for elemental quantification in environmental systems, resulting from its excellent sensitivity and selectivity, which is able to detect specific elements over a broad linear range (Houk et al., 1980). In recent decades, the utility of ICPMS for ENMs has been considerably expanded with the introduction and commercialization of spICPMS (Montaño et al., 2016a; Laborda et al., 2013b, 2011). By introducing intact particles into the plasma and using very short dwell times (10 ms or less), the pulse of ions generated from an ENM ionization event can be detected and quantified as a discrete event above the background. The intensity of the pulse is proportional to the mass of ions generated from the particle events, which assuming a geometry and a material density can be converted into a measure of size (Degueldre and Favarger, 2003; Degueldre et al., 2006). The number of such pulses detected can be related to the number concentration of ENMs in the original sample. In this way, spICPMS is both a particle sizing and a particle counting technique capable of detecting and characterizing ENMs at environmentally relevant concentrations and in complex matrices. Moreover, this technique is ideal for circumventing artifacts generated from sample preparation, as samples can be analyzed directly in their native matrix requiring dilution only when necessary (Montaño et al., 2016a, 2014). However, despite its broad applicability and analytical prowess, spICPMS is still confounded by the presence of naturally occurring nanomaterials. When analyzing materials with low crustal abundance (Au, Ag, and Pt), there are few naturally occurring analogues in natural systems; however, many of the most frequently used and produced ENMs are those comprised of elements (Si, Al, and Fe) that are commonly found in nanominerals and clay particles in environmental systems (Montano et al., 2014).

To navigate this hurdle, ICP-time-of-flight-MS (ICPTOFMS) has been employed as a means of delineating between elementally or isotopically distinct nanomaterial populations (Borovinskaya et al., 2013). As opposed to quadrupole ICPMS, which filters out all ions but those with a specific mass-to-charge ratio, time-of-flight mass analyzers permit all ions into the extraction region where they are then separated and detected according to their mass-to-charge, which is proportional to their time-of-flight (Myers et al., 1994; Dziewatkoski et al., 1996). As a consequence of this configuration, the atomic mass spectra ($\sim$7–250 m/z^+) can be detected and quantified with a time-resolution of 33—46 μs, making it ideal for the characterization of multi-element nanoparticles (Hendriks et al., 2018, 2017). Despite its infancy, this technique has demonstrated considerable promise in the characterization of multielement core-shell particles (Borovinskaya et al., 2014; Naasz et al., 2018). In application to environmental systems, spICPTOFMS can quantify the elemental ratios on a single-particle basis, which has been demonstrated to distinguish between CeO_2 ENMs that contain only Ce and naturally occurring clay particles that often contain both Ce and La in a characteristic ratio (Praetorius et al., 2017). Though obstacles arising from issues in sensitivity and matrix effects are inevitable, this technique represents a promising direction for the future of nanometrology and the study of the nanogeochemical environment.

2.6 Near-infrared fluorescence spectroscopy

Carbonaceous nanomaterials have long presented a challenge in their environmental detection and characterization (Von der Kammer et al., 2012). The abundance of carbon-containing species in the environment ranging from natural organic matter and biogenic molecules necessitates techniques selective toward nanomaterial structures. Due to their nanoscale nature, single-walled carbon nanotubes (SWCNTs) can exhibit chiral-specific, semi-conductive properties, which can be exploited through near-infrared fluorescence (NIRF) (O'connell et al., 2002; Kim et al., 1999). As environmental and biological matrices contain few NIR-absorbing molecules (Von der Kammer et al., 2012; Welsher et al., 2008), this technique enables sensitive and selective determination of SWCNTs down to 1 $\mu g\ L^{-1}$ and approximately 50 $\mu g\ kg^{-1}$ in sediment (Schierz et al., 2014, 2012). When coupled to microscopy, NIRF can be used to examine SWCNT internalization into cells (Welsher et al., 2008) and biological tissues (Bisesi Jr et al., 2014) as a means to track their distribution in organisms and provide a more detailed understanding of their potential toxicity. Despite its sensitivity and selectivity, NIRF is hindered in its environmental application through a number of pathways. The first being that only pristine, semi-conductive SWCNTs may be measured, as metallic SWCNTs will not produce fluorescence. Secondly, aggregated bundles or functionalization of the SWCNTs will quench the fluorescent signals (Petersen et al., 2016; Herrero-Latorre et al., 2015).

2.7 Nanoparticle tracking analysis

Nanoparticle tracking analysis (NTA) is a microscopy technique in which a small volume of the sample is illuminated by an intense laser beam at a 90-degree angle with respect to the objective. The particles scatter light and appear as bright blobs undergoing Brownian motion. The movement of the particles in the sample is monitored at a fixed frame rate using a CCD camera, and the trajectories are tracked during a lapse of time enabling to determine the two-dimensional diffusion coefficients of the particles. The diffusion coefficients are then related to the hydrodynamic diameters of the particles using the Stokes-Einstein relation (Saveyn et al., 2010; Malloy and Carr, 2006). Note that it is possible to detect particles smaller than the diffraction limit. The smallest particle size that can be analyzed is determined by the particles' light scattering power. It is possible to characterize particles as small as 5–10 nm for materials such as Au or Ag that scatter the beam strongly due to a high refractive index contrast between the particles and the dispersion media. Organic matter and silicates might conversely require a diameter up to 30–50 nm to be detectable. Between the smallest detectable diameter and the upper diameter limit of accurate quantification, there is a region where the concentration response, i.e., number of particles visible per frame divided by the number concentration, increases rapidly with size. The measured size distribution of a particle population within this interval might become biased toward larger diameters. This is exemplified by an intermethod comparison study of a sample of c.31 nm silica particles whose mean diameter was by NTA erroneously determined to 68 nm (Tuoriniemi et al., 2014b).

NTA can determine number based size distributions within the diameter range where the technique is suitable with only little bias due to size or material, making it the ideal, if not the only suitable choice for characterizing natural waters where there is a broad continuum of particle sizes of differing materials (Röding et al., 2011). The particle number concentration and size distribution have been measured in a range of environmental water systems including river, lake, sea, and road runoff and wastewater treatment plant effluents (Gallego-Urrea et al., 2010; Sanchís et al., 2015). This data proved to be important for predicting nanoparticles fate in the environment because anthropogenic contaminants will mainly interact (heteroaggregate) with the more abundant background particles (Zanker and Schierz, 2012). However, the ability to distinguish specific particle types in environmental matrices is limited. The scattered light intensity offers limited capabilities for distinguishing between materials and was used for detecting Au particles spiked into natural waters (Gallego-Urrea et al., 2016), fruit juices (Gallego-Urrea et al., 2011), and landfill leachates (Mehrabi et al., 2017). However, the intensity of a light spot also depends on the position of the particle within the intensity profile of the laser beams, wherefore strong scatterers cannot be fully distinguished from the weak ones. Instruments equipped with UV lasers and color filters before the camera have become available and might offer a higher degree of specificity for fluorescent particles, but this capability has not yet been explored for environmental analysis.

The technique has the potential to be improved by analyzing detailed trajectories of the particles (three dimensional) and increasing light power (with the inconvenience of interference of the movement, e.g., optical tweezers) or detector capabilities (filters of emitted light, e.g., fluorescence or multispectral detectors) to investigate more properties from the scattered radiation.

3 Aquatic compartments

3.1 Expected concentrations

As mentioned in previous sections, most of the analytical techniques used at the moment for monitoring ENMs in aquatic systems have been optimized for aquatic tests in laboratory conditions, e.g., ecotoxicological and environmental fate studies (Hassellöv et al., 2008). However, environmental monitoring in natural waters imposes additional requirements on the analytical task (Zanker and Schierz, 2012). Aquatic compartments are dilute and dynamic systems containing dissolved organic matter (DOM), suspended particulate matter (SPM) and inorganic ions at concentrations that vary as a function of space and time (Fig. 2). Freshwater systems, such as rivers, lakes, and groundwaters are typically rich in DOM and SPM with low ionic strength (salinity). On the contrary, sea and ocean waters contain very high salinity

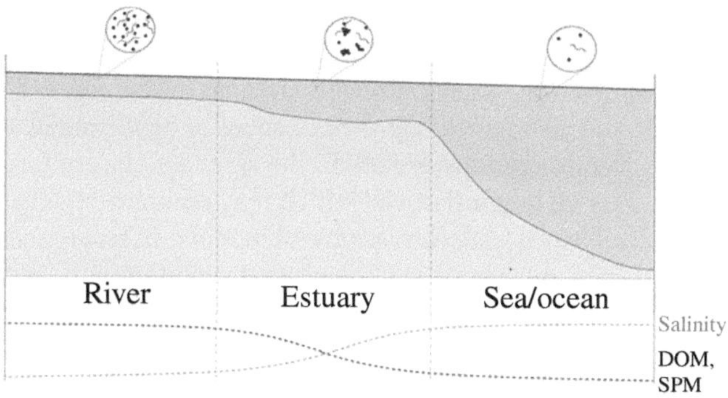

Fig. 2 Graphic representation of the increase in salinity and decrease in dissolved organic matter (DOM) and suspended particulate matter (SPM) of water as it flows through rivers, into estuaries and the ocean or sea.

and low SPM, with seasonal variations of DOM and SPM types and concentration (Linders et al., 2018; Zetsche and Ploug, 2015). Seawater also contains high concentrations of divalent ions, which are effective destabilizers of particles, i.e., they promote aggregation by suppressing the stabilizing effect of the particle's surface charge or by forming bridges between the DOM molecules attached on the particles' surface (Narkis and Rebhun, 1975). Freshwater from rivers meets the ocean seawater in estuary systems where complex and dynamic changes of the water composition occur. The majority of the SPM, including ENMs, is removed from the water column through destabilization and settling on the sediments (Benoit et al., 1994; Stolpe and Hassellöv, 2007). Overall, SPM, DOM, and salinity act as regulators of the environmental fate of ENMs and often dictate the analytical methods used for ENM detection.

ENMs may be released into aquatic systems through wastewater treatment plant effluents, accidental spills of industrial products, and normal use of nano-enabled products (e.g., sunscreen, cosmetics, and ship paint) (Sun et al., 2014; Gottschalk and Nowack, 2011; Arvidsson et al., 2012). Once released, ENMs are diluted into large and dynamic waters, such as rivers, lakes, and seas and may be transported over long distances through advection or diffusion. Recent studies estimate the environmental concentrations of ENMs in sub $\mu g\ L^{-1}$ values (Wang and Nowack, 2018), much lower than in the consumer products from where they were released and lower than in other environmental compartments such as soils (see further), which complicates their detection in natural aquatic environments. Quantifying total mass concentrations of ENMs in aquatic compartments thus requires particularly low detection limits. The quantification of ENM mass concentrations is also complicated, but less so than terrestrial compartments, by the presence of naturally occurring particles in the nanometer range making it difficult to identify target elements material with common screening techniques, such as NTA or DLS that detect particles non-specifically.

Once in the environment, various transformation processes may lead to homoaggregation (aggregation between ENMs), heteroaggregation (aggregation between ENMs and SPM), and various surface transformations (Nowack et al., 2012; Levard et al., 2012). In order to characterize the many possible forms of occurrence of ENMs comprehensively, both form (size and shape) and composition (target elements, structure, and crystallinity) need to be determined. Homoaggregates contain elevated concentrations of the target element compared to individual ENMs, making their detection easier, but also increasing their chances to be mistaken as larger solids (not nanomaterials). Heteroaggregates contain the same amount of target element as single ENMs, but in a much larger particle, which may complicate detection (Von der Kammer et al., 2012). Surface transformations include redox reactions, adsorption of compounds, desorption of the engineered coating, dissolution, etc. When such reactions are favored, time resolved measurements of dissolved compounds involved in the reaction is often used to dynamically track the evolution of ENM transformation (Gondikas et al., 2010, 2012). This can be achieved either using ultrafiltration (Cornelis et al., 2010), dialysis (Franklin et al., 2007), or ultracentrifugation (Liu and Hurt, 2010), each method having its advantages and disadvantages (Utembe et al., 2015).

The ideal set of measurands to obtain full information on the ENMs form of occurrence includes number concentration, particle composition, size, coating, and shape, among others (Hassellöv and Kaegi, 2010). In most cases, several complementary techniques need to be combined in order to face the dynamic state and complexity of ENMs in the aquatic compartment. In addition, sample treatment (e.g., mild acid digestion) or separation techniques (e.g., fractionation, or filtration) as described for terrestrial compartments can be problematic due to the low concentrations found and the large amount of particulate matter present (Zanker and Schierz, 2012). For example, mild digestion of organic particles may help to keep the original shape and form of inorganic particles while strong digestion may only be used to detect highly refractory nanomaterials such as TiO_2. Preconcentration and filtration although effective, may lead to substantial loss of target ENMs on the membrane surface (Buesseler et al., 1996). FFF greatly simplifies the detection of ENMs, however it is hampered by the tedious and time-consuming method development process.

3.2 Single particle ICPMS

spICPMS is one of the main techniques used for identifying metal-based nanoparticles in aquatic systems, because it offers the lowest detection limits in terms of concentration to date. Moreover, the technique allows the measurement of the target element in single particles, which is used to derive the number concentration and size distribution of these particles (Tuoriniemi et al., 2014a; Gondikas et al., 2014; Pace et al., 2012). The method is fairly straightforward to use; however, sample characteristics may necessitate technical adjustments and care should be taken in the interpretation of the data. Measuring extremely low number concentrations using spICPMS is in theory only limited by measurement time. spICPMS is a counting technique, so better statistics and, thus, more accurate counting values are obtained by extending the analysis time (Tuoriniemi et al., 2012). The low expected concentrations of ENMs in the aquatic compartment would indeed require lengthy measurement times and analysis of multiple replicates (Gondikas et al., 2018). Note also, that the dynamic range in terms of number concentrations is limited to ca. three orders of magnitude, because multiple particle events outweigh single particle events (Montaño et al., 2016a; Tuoriniemi et al., 2012). Operating spICPMS in the so-called fast-acquisition speed technology (FAST) extends this dynamic range because multiple particle events become much less frequent (Montaño et al., 2014). Furthermore, light elements may be difficult to quantify and it is not possible to measure carbon.

There are both lower and upper limits to the dynamic range for spICPMS in terms of size. The lower size detection limit depends on the ionization potential of the element that is monitored as well as the dissolved background of that element (Lee et al., 2014). An upper detection limit is imposed by the spray chamber that does not let large heteroaggregates (> ca. 10 μm) in the ICPMS. While such particles are beyond the range of what is considered an ENM, large heteroaggregates are indeed effectively undetectable using spICPMS (Tuoriniemi et al., 2017; Winterwerp, 1998). Large, solid metal ENMs are also analyzed less accurately because the dead-time correction no longer suffices to correctly analyze the extremely intense pulses generated by such particles (Liu et al., 2014; Lee and Chan, 2015). Depending on the nebulizer and spray chamber used, particles larger than approximately 10 μm may thus need to be removed prior to analysis, by means of filtration or centrifugation.

Baseline signal drift is a known problem during spICPMS measurements, when the sample contains high SPM or DOM concentrations or a high salinity. Given that most ICPMS instruments can only measure one element at a time in single-particle mode, regular sensitivity checks need to be done (Tuoriniemi et al., 2017). Moreover, the operator needs to conduct frequent maintenance on the instrument and care should be taken that the total suspended solids do not compromise plasma stability. Samples with high SPM are therefore typically diluted prior to analysis, but dilution further extends the duration of measurements in order to obtain a statistically significant particle count. Note also that dilution may incur changes in particle stability, e.g., the drop of salinity may result in disaggregation or alternatively, a dilution in stabilizing DOC may induce aggregation. For that reason, it is preferable that the time between the dilution and measurement is minimized (Toncelli et al., 2016). The presence of other particulate material, transformation processes, and variability in water composition may call for use of spICPMS hyphenated with other methods for monitoring natural waters (Gimbert et al., 2005). One example is using FFF as a separation technique prior to detection with spICPMS (e.g., Huynh et al., 2016). This system combines the superior capabilities of FFF for determination of particle size distributions with the advantages of spICPMS, and It is one of the best options available, in terms of quality of information acquired. However, the method development is tedious and the measurements duration is lengthy. For those reasons, it is unlikely that this combination is used for the routine analysis. Application of the spICPTOFMS technique in model samples has shown that multielement fingerprints may be identified on single particles and thus identify the ENMs origin (Praetorius et al., 2017). The technique maintains the simplicity of sample treatment of spICPMS, but provides a more holistic picture on the composition of single particles, which substantially improves the estimation of particle size. Carbon is still not measurable; data treatment is much more complicated, but the method has been applied in environmental samples with great promise (Gondikas et al., 2018).

3.3 Electron microscopy and light scattering

Electron microscopy is a versatile technique that provides high quality information on form and composition of single particles and aggregates. Combined with the possibility to be used with multiple detectors (e.g., EDS and Raman), the information obtained is superior to any other technique. However, these advantages are applicable only after ENMs are detected on the electron microscope, which is not a trivial task. Firstly, particles contained in a water sample need to be isolated from the water, either by means of ultracentrifugation or filtration with very narrow pore size membranes (smaller than a few nanometers), followed by evaporation of the remaining water (Mavrocordatos et al., 2007). This separation step is very likely to change the physical state of the particles, e.g., aggregation induced during drying. This obstacle may be partially overcome by the use of stabilizing agents prior to drying (Michen et al., 2015) or with the use of cryo-TEM. However, tedious method development is necessary to make sure there are no complications induced by the unknown composition of natural water samples. Secondly, scanning the sample to detect nanoparticles is a time-consuming process, subject to operator bias. For those reasons, electron microscopy is not likely to be used for routine measurements but may be used to analyze selected samples for verification purposes.

In the industry, the most widely used techniques for measuring particle size are based on light scattering. Some of these techniques rely on averaged intensity of particles, which requires relatively high concentrations to provide reliable data (Gallego-Urrea et al., 2014). This makes techniques such as DLS not useful for monitoring natural waters, but there are a few promising alternatives. NTA is a single particle method with much lower limits of detection compared to DLS, but as with all light scattering techniques, it is unable to distinguish between natural and anthropogenic particles. Therefore, it may only be considered as a screening technique to provide information about the unusual presence of particulate material in the nanorange, which might call for further analysis, e.g., in drinking water facilities. Centrifugal techniques that use light absorption or transmittance as detectors such as disc centrifuge or centrifugal separation analysis can also screen for ENM analysis. However, none of these techniques are able to identify the elemental composition of particles; thus, they may only be used for initial screening purposes.

3.4 Detection of ENMs in the aquatic environment

The use of ENMs in consumer products is currently in a premature stage, mostly due to the uncertainty that surrounds future legislation. As a result, emission of ENMs in the environment is currently limited, but it is expected to intensify in the future, when nanoenabled products become established in the market. This poses a challenge for testing analytical methods in natural systems. There are only a few such attempts, which unavoidably target ENMs with the highest production volumes, i.e., titanium dioxide and silicon dioxide. However, these materials are abundant in nature and are frequently found as natural nanoparticles (Wagner et al., 2014). It has become apparent in these test cases that distinguishing between the anthropogenic and natural nanomaterials is essential for determining release sources and exposure levels.

Gondikas et al. studied TiO2 ENMs released from sunscreen during bathing activities in a lake (Gondikas et al., 2014, 2018). They used elemental ratios of the bulk SPM, sediments, and soil, as well as spICPMS, spICPTOFMS, and TEM on lake water samples to determine that Ti elemental concentrations were consistently higher in single particles than in the bulk samples. Interestingly, an accumulation of TiO_2 nanoparticles was found on the surface of the lake, i.e., the top few centimeters of the water, which is most exposed to sunlight and may trigger redox reactions. Reed et al. (2017) combined spICPMS data with measurements of organic compounds used in sunscreen to determine the input of TiO2 ENMs in river water. They revealed that in shallow water bodies, a large part of TiO_2 increase in the water is caused by sediment resuspension, from bathing activities. In a more recent study, Loosli et al. (2019a)used a combination of spICPTOFMS, bulk analysis, and TEM to determine TiO_2 ENMs released from sewage spills. They found evidence that natural TiO_2 ENMs often contain a group of elements, Al, Fe, Nb, and Pb, among others. Elemental ratios of Ti with these elements on single particles combined with particle size and shape determination is the current state-of-the-art in monitoring of ENMs in natural waters. Improvements on the analytical instrumentation is continuous, even for the most difficult elements, such as SiO_2 (Montaño et al., 2016b), so it is likely that routine monitoring of metal-containing ENMs will be achieved in the near future.

Not only metal-containing ENMs are released in the environment from anthropogenic activities. Nanoplastics are a group of ENMs that have recently received attention due to the omnipresent pollution caused by plastics. However, the definition of nanoplastics in the current literature is different from that of ENMs, because the former is often defined as plastic particles smaller than 1 μm (Gigault et al., 2018), while ENMs are often defined as materials that are smaller than 0.1 μm in at least one dimension. Nanoplastics are most likely present in the aquatic environment because larger pieces of plastic fragment and degrade, and there is no reason to believe that these particles will not degrade into the nanometer size range (Horton et al., 2017). However, there are no data available regarding detection and identification of truly nanosized plastic particles (one dimension less than 100 nm) in the aquatic environment. The reason is that most sampling and analysis techniques do not target nanoplastics. Three of the most widely used methods for sampling plastics in the aquatic environment can in principle sample nanoparticles: surface water sampling, biological sampling, and sediment sampling (Hidalgo-Ruz et al., 2012). Sampling nanosized plastic particles from surface water is difficult because the amount of nanoplastic particles in relation to organic matter is low; thereby, a large volume is needed with as little organic matter as possible. For sediment samples, a density separation method is applied to separate the plastic particles from the sediment. The lighter plastic particles float and are filtered through filters with different pore sizes, and in theory, it would be possible to use filters with smaller pore size, such as 20 nm to collect nanoplastics. However, other particles can also end up on the filters. With biological sampling, if nanoplastics are present in the organism, these particles are collected. However, the main challenge is currently to find them and identify them.

Raman spectroscopy or FTIR, often in combination with microscopy, is currently the most used technique for detecting nanoplastics, e.g., in the marine environment, because it is the only approach by which both size and composition information can be

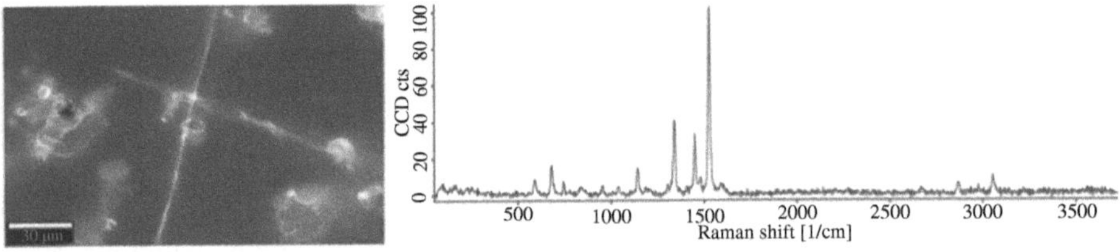

Fig. 3 Detection of nanoplastic particles in seawater using Raman-microscopy. Left is the microscopy image and the right is the Raman spectrum that can be used to identify the nanoplastic particle shown.

obtained for nanoplastics (Lenz et al., 2015; Song et al., 2015; Zhao et al., 2018) (see also Fig. 3). Pyrolysis or TGA-GC-MS have used, but the sample needs to be filtered and dried for these techniques (Dümichen et al., 2017, 2015). In the latter case, no-size information is, of course, obtained.

4 Soils and sediments

4.1 Expected concentrations

Terrestrial compartments such as soils and sediments consist of diverse aggregated particulate matter. Both size and chemical specificity is thus a requirement for ENM monitoring. Methods for trace particle analysis are considerably less well developed compared to aqueous matrices, but the concentrations of ENMs are predicted to be orders of magnitude higher. Examples of predicted concentrations of ENMs in European soils are ca. 1000 μg kg^{-1}, 0.01 μg kg^{-1}, and 0.1 μg kg^{-1} for TiO_2, ZnO, and Ag respectively, and the corresponding concentrations of these materials in sediments are ca. 1000 μg kg^{-1}, 100 μg kg^{-1}, and 1 μg kg^{-1} (Sun et al., 2014). The total concentrations including all particle sizes of these materials are several orders of magnitude higher, so total elemental concentrations can hardly be used as a proxy for NM. Soils and sediments are a sink for ENMs, and models predict the concentrations in soils to increase by 0.24 μg kg^{-1}, 0.03 μg kg^{-1}, and 0.0018 μg kg^{-1} per year, while the increase in sediments are 4800 μg kg^{-1}, 1000 μg kg^{-1}, and 3.3 μg kg^{-1} for the TiO_2, ZnO, and Ag respectively (Sun et al., 2014). Waste water sludge treated soils receive orders of magnitude higher ENM doses while at the same time being used for food production. Other important key environmental compartments for which the considerations in this chapter apply are dried waste water sludges, fly ashes, and highway dusts.

4.2 ENM extraction

Most soil analyses involve prior extraction, because the most powerful techniques for ENM analysis use ICPMS analysis, which requires aqueous dispersions. Any extraction procedure must therefore provide ENMs that do not reside on aggregates >1 μm because there is a risk that such big particles are lost in the spray chamber and/or block the nebulizer (Tuoriniemi et al., 2017). Also, the total suspended solids content that can be introduced into a plasma is limited to 1%–2% (International Organization for Standardization, 2009). Nebulizing a slurry of milled soil dispersed in water is therefore not likely to be practical. Contrary to a total element determination, powerful agents such as concentrated acids cannot be used because these often dissolve the ENMs preventing their detection. $Na_4P_4O_7$ solutions have proven to be a mild extraction agent that still extracts a considerable fraction of the total concentration of inorganic nanomaterials from soils, usually aided using ultrasonication (Schwertfeger et al., 2017; Loosli et al., 2019b). More recalcitrant ENMs such as TiO2 can be extracted using stronger reagents such as a H_2O_2/H_2SO_4 mixture (Philippe et al., 2018). The low specific density of nanoplastics allows the use of flotation methods (Fries et al., 2013). With all extraction methods, however, only a fraction of the total ENM concentration can be extracted (Schwertfeger et al., 2017; Loosli et al., 2019b), and it is unclear what this fraction really represents and how representative it is for the total ENM population and its original form of occurrence prior to extraction. This is, however, a known problem for the case for conventional element determination in soils where concentrations measured are defined by the digestion procedure used (Gupta and Aten, 1993). The ENM speciation in soil, i.e., binding to specific mineral and size fractions in the soil is likely to affect the extraction efficiency. Therefore, a recovery of particle standards spiked immediately before a method development experiment might not be representative for particles that have resided for years in a given sample.

Techniques that analyze soils without sample pretreatment are therefore preferable. There are emerging counting techniques based on SEM and spICPMS although these are far from being ready for routine monitoring applications. An emerging technique for carbon-based ENMs such as nanoplastics is Py-GC-MS (Dümichen et al., 2017, 2015). By itself, it only measures concentrations without giving any information on the particle size distribution though it could be applied on size fractions isolated by FFF or hydrodynamic chromatography, in both cases following extraction. No size distributions of nanoplastics have hitherto been measured in any type of environmental sample, let alone in soils or sediments.

4.3 SEM-BSE analysis

Inorganic ENMs can be identified, quantified, and sized in soils using SEM-BSE (Tuoriniemi et al., 2014c). With BSE, the contrast is mainly atomic number-based, with ENMs containing relatively electron-rich elements appearing bright against the dark background of relatively electron-poor silicates and organic matter (Fig. 4). The elemental composition can be determined using EDS or WDS. This imaging technique forms the basis of automated SEM systems such as QEMSCAN that are commonly used in petrography and forensic trace particle analysis (Pirrie et al., 2009). BSE-SEM has also been applied to measure the ENM number concentration per mass of soil together with the size distribution, but some further developments are required. It is possible to determine the number concentration from the number of particles visible per imaged area, provided that the probing depth at which

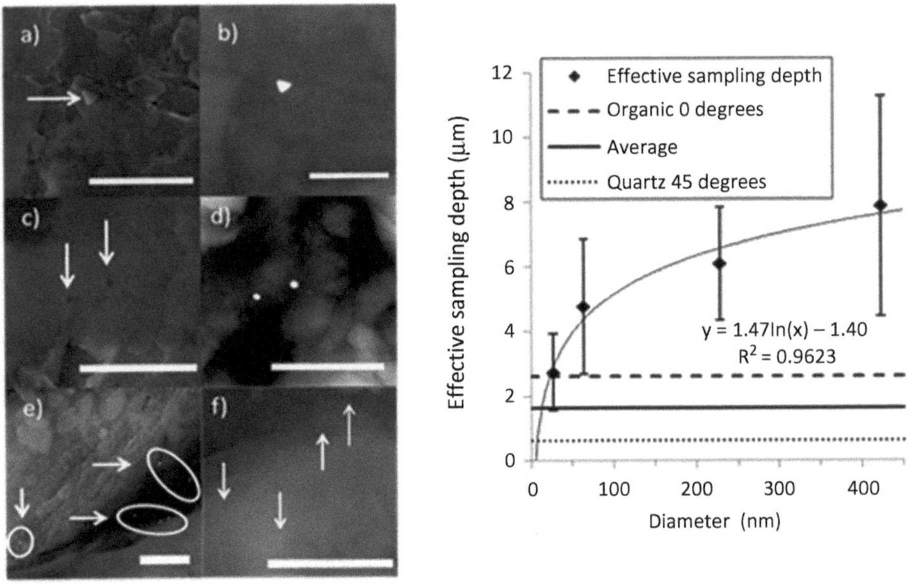

Fig. 4 Left side: Representative images of Au ENMs spiked into soil acquired with a FEI Quanta 200 ESEM equipped with a FEG and operated in the low vacuum mode at a pressure of 0.5–0.6 Torr with an acceleration voltage of 20 kV and a working distance of 10 mm. The scale bar is 2 μm in all images. (A and B) SE and BSE images of a 250 nm particle. (C and D) SE and BSE images of two 100 nm particles. (E and F) BSE images of 50 nm and 30 nm particles respectively. The arrows indicate particles. Right side: Effective values of the sampling or probing depths as a function of Au ENM size measured by counting the particles per imaged area for known number concentrations spiked into the soil. The error bars are the 95% confidence interval due to variation between the experiments. A logarithmic function is fitted to the data. The horizontal lines show the probing depth calculated from the Kanaya-Okayama electron range for different soil components and electron beam incidence angles.
Reprinted with permission from Tuoriniemi, J., et al., 2014. In situ characterisation of physicochemical state and concentration of nanoparticles in soil ecotoxicity studies using environmental scanning electron microscopy. Environ. Chem. 11 (4), 367–376.

ENMs can be detected is known (Tuoriniemi et al., 2014c). Unfortunately, the probing depth for ENMs is both size- and material-dependent, because larger and denser particles scatter a larger fraction of the incident electron beam.

Differently sized Au ENMs were spiked into soils at different number concentrations, and the BSE-visualized ENMs were counted in a given area (Tuoriniemi et al., 2014c); it was found that the probing depth increased logarithmically with the ENM size, which is theoretically expected. However, the electron beam spreads out upon entering the soil making it difficult to locate edges of ENMs, and larger particles are more frequently detected due to the deeper probing depth. The size of Au ENMs having a certified SEM diameter of c.27 nm was thus quantified as 38 nm when immersed in soil (Tuoriniemi et al., 2014c). This sizing bias clearly needs more systematic investigation.

The roughness of the soil surface exposed to the electron beam, however, also has a strong effect on the probing depth and, thus, the quantified number concentrations. The Kanaya-Okayama electron range in soil was estimated to c.2 μm while the probing depth reached ca. 7 μm for the larger Au ENMs. This indicates that surface roughness increases the sample area exposed to the electron beam (Tuoriniemi et al., 2014c). A possible improvement is to mill the soil and deposit it as a thin (<2 μm) homogeneous film to be uniformly penetrated by the electron beam. It could then be possible to calculate an estimate of the probing depth for a given SEM acceleration voltage by Monte Carlo simulations of electron trajectories in the material (Österreicher et al., 2018). This has only been proven to work for matrices consisting of relatively homogenous alloys. The approach is likely complicated for an actual soil matrix that is heterogeneous on the scale of the electron range, and neither the composition nor relevant materials properties of the soil components are known accurately enough to do Monte Carlo modeling. Probing depths therefore have to be determined by calibration using particles of known size and number concentration, preferably having the same composition as the unknown particles. While such standards exist for ENMs such as Au or Ag, these standard particles are only certified for their size, and reliable, stable ENMS suspensions with a certified number concentration do not yet exist.

A remaining problem this application shares with all microscopy approaches is the poor number concentration detection limit. It can be calculated that an area corresponding to a few hundred images would have to be scanned to detect a single 50 nm Ag ENM in a sample that contains 1 μg/L Ag ENMs. Analysis of realistic, sub mg L^{-1} concentrations is therefore time-consuming and thus expensive even with an automated system where parameters such as beam scan speed and spot size would be optimized for rapid scanning of large areas and detection of BSE signal spikes rather than image quality.

4.4 Laser ablation single particle ICPMS

Elemental contents of solid samples can be directly determined with laser ablation (LA) ICPMS without dissolution of the sample. Solid material is ablated from the sample by a high energy laser beam. The ablated material is transported with a gas flow into the ICPMS detector where elemental masses are quantified. The technique can also obtain spatially resolved data when xy-coordinate data of the laser is related to the time-resolved ICPMS analysis (Becker et al., 2007). Spatial resolution, however, is above single nanoparticle size. Currently, spot sizes of approximately 5 μm can be accomplished in practice (Gundlach-Graham et al., 2018; Arakawa et al., 2019). Quantification of the elemental content is done by using matrix-matched standard materials. For tissue samples, agarose gel has been suggested as a matrix matched standard (Stark and Wennrich, 2011). LA-ICPMS has been implemented in the field of nanometrology, and the distribution of ENMs in tissues was quantified (Bohme et al., 2015). It has been observed that particles may not be destroyed during the ablation process but only be mobilized from the solid sample (Benesova et al., 2016).

This may allow to obtain not only elemental mass data but also single particle size information.

Yet unpublished research by the authors has shown that superficially attached ENMs are indeed mostly transferred intact from soil into an ICPMS plasma by LA. The ENMs could thus potentially be counted and sized using spICPMS in much the same way as this technique is used for aqueous samples, provided that the ablated soil quantity can be quantified accurately. During a LA-spICPMS experiment, the soil is first manually homogenized, and a known soil mass is spread on a double-sided tape. The tape piece is thus coated with a ca. 10–15 μm thick layer of soil with a known average adsorbed mass per tape area. It was found that the soil layer can be ablated into the plasma with a recovery of around 70%–80%, and the associated ENMs appear as spikes in an otherwise continuous background signal.

One has to compensate for the ionization suppression effect of the soil fragments when calibrating the mass spectrometer sensitivity. This was achieved by having the plasma connected to both the LA cell and a nebulizer-spray chamber system simultaneously using a T-section. ENMs for nebulization efficiency determination and dissolved elemental standards could thus also be injected into the plasma while ablating the sample for standard addition calibration. The sizes of 60, 100, and 250 nm nominal diameter Au ENMs spiked into soil were correctly reproduced by LA-spICPMS. The particle event frequency was linearly dependent on the spiked particle concentration in the soil.

The size detection limit of LA-spICPMS is limited by instrument sensitivity, dissolved background/carry-over of the element of interest, and the dwell time, similar to aqueous samples. An additional complication of LA-spICPMS, however, is that all dissolved ions appear as particle spikes, because the ions are adsorbed to ablated fragments carried into the plasma. All dissolved ions on one fragment thus arrive roughly at the same time at the detector creating an ENM-like spike. This is not unlike aqueous spICPMS where dissolved ions eventually become nanoparticles during evaporation of the solvent, e.g., in the plasma or membrane desolvator, but it remains to be investigated whether the phenomenon is more pronounced during LA-spICPMS.

The number concentration could be quantified either by ablating a known area with a certain mass of soil on top of it and assuming a full recovery of the particles, or one could measure the ablation rate, which is the mass of soil reaching the plasma per second, using a soil with a known concentration of a metal (e.g., In) spiked into it. The latter method appeared more accurate. A typical ablation rate is 10^{-8}–10^{-7} g s^{-1}. A given volume of soil is therefore scanned for particles 10^3–10^4 times faster than with SEM, while both the concentration and size determination are more accurate. There is thus a potential to reach much lower number concentration detection limits in a realistic amount of time at a much lower cost compared to SEM. A future development would be to combine LA with the multielement capabilities of TOF-ICPMS instruments. Similarly, in aqueous samples, a broader range of ENMs could thus be distinguished from naturally occurring inorganic particles, provided that machine learning algorithms are developed (Praetorius et al., 2017).

5 Wastewater treatment sludge and other solid samples

5.1 Expected concentrations

One of the aims of wastewater treatment is the removal of particulate matter from the wastewater. Therefore, in wastewater treatment plants (WWTPs) particulate matter is efficiently accumulated in the sludge and only a small mass fraction of the particulate matter is released into surface waters by the treated wastewater effluent (Sun et al., 2016, 2014; Park et al., 2013; Kaegi et al., 2013). Hence, ENM constituents as part of the particulate matter are also accumulated in the sewage sludge or more precisely in the activated sludge. The activated sludge is further treated, e. g., by anaerobic digestion, dewatering, or incineration forming solids with different properties.

ENM concentrations in the sludge may be predicted based on material flow models or on removal rates of ENMs from wastewater in WWTPs. These data indicate elevated concentration levels compared to surface water as well as wastewater effluent. Bulk analysis of total elemental concentration after chemical digestion of sludge samples may give an indication of the presence of specific ENMs, whereas information on particle size as well as information on elemental speciation will be lost. With this approach, it is also not possible to differentiate between the various types of ENMs, matrix particles, and dissolved constituents containing similar elements. As a preparatory step, total digestion aims to dissolve all particulate matter including the ENMs of interest and to determine the resulting dissolved elemental concentration. Digestion of the particulate matrix is commonly achieved by the addition of an oxidizing agent and possibly an acid or base. To enhance digestion, elevated temperature and pressure may help, which can be provided by microwave assistance. Combinations of H_2O_2 and HNO_3 are often deployed to oxidize organic carbon and dissolve, e.g., Ag ENMs. SiO_2 and TiO_2 ENMs will retain this treatment and can be dissolved by other reagents, e.g., by hydrofluoric acid digestion (Wagner et al., 2015).

5.2 Challenges

To elucidate fate processes of ENMs in WWTPs and to determine total ENM contents, analyses of the sludge and its solid products are required. Up to now, there are no routine monitoring data available on ENM concentrations in sewage sludge based on actual measurements (Bundschuh et al., 2018). This is due to a couple of still existing analytical challenges related to conditions in sewage sludge: high particulate matrix concentration with high organic carbon content, the presence of numerous types of various ENMs, matrix particles of similar elemental contents, and the dynamics in chemical conditions (aerobic to anaerobic) as well as the high biological activity in the particulate matrix. These characteristics have a number of implications for ENM fate and analyses: (i) transformation of redox sensitive species is expected and special care thus needs to be taken during sampling, (ii) attachment of ENMs to other particulate matter (heteroaggregation) takes place, (iii) ENM surfaces are coated with dissolved organic matter (Kaegi et al., 2013) and (iv) particles of similar chemical identity but of different sources are present. To overcome these implications, methods for ENM identification, quantification and characterization are still under development (Laborda et al., 2016). These analyses of ENMs in sewage sludge and other solid samples address mostly the following measures: total ENM content and characterization of the ENM in the solid regarding particle size, morphology, chemical composition, chemical speciation, and crystallinity.

5.3 ENM extraction

There are methods being developed (cryo-ToF-SIMS and cryo-EM) which may be able to characterize ENMs under in situ conditions (Benettoni et al., 2019). Ag ENMs attached to fresh sludge was also injected directly into an ICPMS allowing to analyze the size distribution as a function of time (Tuoriniemi et al., 2017). This was only possible for a limited number of samples, however, as deposition of carbon of the WWTP sludge matrix continuously narrowed the ICP cones causing massive drift that needed to be corrected continuously. Moreover, large WWTP flocs were not able to pass the spray chamber of the ICPMS, giving an incomplete information on the Ag ENM form of occurrence. ENM concentration and characterization in activated sludge or other solid samples such as ashes can therefore in most cases only be determined after extraction of the particles of interest (Von der Kammer et al., 2012). Similarly, in soils, however, extraction methods may struggle with incomplete extraction of the ENMs and changes of the ENM properties (Laborda et al., 2016). Therefore, many studies addressed the optimization of the ENMs extraction from solid matrices (e.g., Schwertfeger et al., 2017; Loosli et al., 2019b).

The matrix of WWTP is centered between soils and aqueous samples in terms of complexity. It is much less dense than that of soils, but a mere filtration cannot be used to separate ENMs from WWTP sludges due to the presence of similar sized particles with different chemical identity and associations between ENMs and other particles. Density gradient centrifugation may conversely be applied with some success to separate inorganic ENMs that have a higher density compared to the bulk of the background matrix. The matrix of WWTP is also easily removed compared to soils, and this approach may be applied if ENMs are resistant to the digestion technique that dissolves the matrix. TiO_2, for instance, does not dissolve during H_2O_2/HNO_3 treatment that destroys the organic matrix components. As a complete separation of ENMs from the matrix is almost impossible during the extraction procedure (Laborda et al., 2016), field- flow fractionation and hydrodynamic chromatography techniques have been developed, which serve as a particle size separation prior to analysis. These methods allow to separate a complex particle suspension, for instance, after extraction according to hydrodynamic particle size (Kaegi et al., 2013; Ma et al., 2014). The size fractionated sample as well as the extracts may be further analyzed using spICPMS

(Philippe et al., 2018; Rakcheev et al., 2013). spICPMS analysis reveals information on elemental concentration associated with particulate material and particles size.

5.4 SEM/TEM

There are a number of studies addressing the fate of ENMs in wastewater treatment facilities, specifically in sewage sludge and sewage sludge treatment based on spiking experiments (Kaegi et al., 2013; Tiede et al., 2010; Meier et al., 2016). Such investigations rely on characterization of ENMs in solids like activated sewage sludge or in biosolids after anaerobic digestion. Characterization of ENMs in the solid refers mostly to particle size, morphology, and also crystallinity and elemental speciation. Particle size can be determined after extraction by SEM or TEM analysis or by single particle ICPMS analysis (Meier et al., 2016).

The fate of Ag ENMs spiked to sewage sludge was investigated by SEM and TEM analyses revealing, for instance that heteroaggregates rather than homoaggregates occur between Ag ENMs and organic particulate matter (Kaegi et al., 2013). Due to the dynamics of redox conditions during the processing of wastewater treatment sludge, redox sensitive species may change. To track these changes SEM-EDX as well as X-ray absorption spectroscopy (XAS) may be used. Transformation of Ag ENMs into Ag_2S particles within hours was first reported based on SEM-EDX analysis (Kaegi et al., 2013). XAS was used to determine the chemical speciation of Ag and ZnO nanoparticles after anaerobic digestion (Ma et al., 2014) or in activated sludge (Kaegi et al., 2013). These studies showed that metallic Ag ENMs are transformed into Ag_2S particles. Zn was transformed to ZnS, $Zn_3(PO_4)_2$ and Zn associated with Fe oxy/hydroxide depending on the redox state and water content. Data comparison between nanoparticlulate and dissolved forms revealed that both behave similarly, and nanoparticles transform to similar chemical forms as bulk metals in the WWTP (Ma et al., 2014). Sludge after anaerobic digestion is incinerated or used as a fertilizer. During incineration Ag0 forms rapidly from Ag_2S (Meier et al., 2016), which was demonstrated by XAS and SEM-EDX analyses.

5.5 Laser ablation single particle ICPMS

A more direct approach, not requiring any prior extraction or size separation may be to use LA-spICPMS, similarly as for soil samples. It was shown by the authors that LA-spICPMS can also be used for detecting Ag, Au, and CeO_2 ENMs in wastewater treatment sludge, fly ash, and road dust (to date unpublished results). The WWTP sludges and other solid samples were dried and superficially applied to double-sided tape, and similar recovery rates were obtained as in the soil experiments. It was found, however, that calibration must occur while matrix matching the ablated matrix of sludges. This somewhat limits the environmental monitoring of unknown samples, i.e., non-spiked samples, because a reference matrix representative for the unknown sample needs to be obtained. This is even more problematic for detecting ENMs such as CeO_2 that occur ubiquitously in WWTP sludges and other solid samples, albeit in different forms of occurrence than purposely made CeO2 ENMs. A LA-spICPMS analysis does not allow differentiating between particles of anthropogenic origin and natural occurring counterparts with similar elemental content. An analytical approach to resolve identification of ENMs has been described earlier, which generally focused on the single particle analysis using ICP-ToF-MS (Praetorius et al., 2017) and could also be applied in this case.

6 Conclusion

Table 1 summarizes some of the analytical capabilities of the different techniques discussed. Recent developments, especially in terms of spTOFICPMS have significantly expanded our analytical capabilities for environmental monitoring of inorganic ENMs. This technique can measure ENM mass concentrations at realistic environmental concentrations in aquatic systems and even distinguish some major forms of occurrence (Fig. 1) with relatively little prior sample treatment. Future, more complex ENMs will likely still be able to be analyzed given the multielement capability of this technique. It has, however, not yet been explored whether the technique can be coupled to LA and still retain the same analytical capability when analyzing solid samples.

The capability for measuring carbon-based ENMs clearly lags, both in terms of selectivity and detection limits, but particularly in terms of size capabilities. Given that carbon-based ENMs such as graphene, carbon nanotubes, and nanoplastics have become increasingly of concern, it is especially in these areas that developments are required the most. Advanced hybrid ENMs will likely contain a carbon-based portion, even if only as a coating, and this portion may provide very important functionalities that requires monitoring in environmental compartments.

Acknowledgments

Geert Cornelis, Jani Tuoriniemi and Stephan Wagner were supported by the European Union's Horizon 2020 research and innovation programme under grant agreement 720952 (Acenano). Geert Cornelis, Jani Tuoriniemi and Julian Gallego-Urrea were supported by the European Union's Horizon 2020 research and innovation programme under grant agreement 646002 (Nanofase). Stephan Wagner was supported by the Helmholtz Research Program Earth & Environment, thematic area Chemical in the Environment. Julian Gallego-Urrea was supported by the FRAM Centre for Future Chemical Risk Assessment and Management Strategies at the University of Gothenburg. Andreas Gondikas has received funding from the Hellenic Foundation for Research and Innovation (HFRI) and the General Secretariat for Research and Technology (GSRT), under grant agreement No 1317 (Nano Island).

Table 1 Overview of capabilities of different techniques for environmental monitoring.

Technique	Aqueous/solid samples[a]	Inorganic/ C-based	Most suited for	Interferents	Concentration LOD[b] (μg/L)	Size LOD[b] (nm)	Number concentration	Sub-techniques and hyphenations
TEM	All	Both	Electron-rich ENMs		100	1 nm	Yes	EDX, EELS
SEM	All	Inorganic	Electron-rich ENMs		100	5 nm	Yes	EDX, BSE, Raman
FFF-ICPMS	Aqueous	Both[c]	heavy elements	Particles having the same elements in the same ratio	10	1 nm	No	Raman, DLS, SLS, spICPMS
IR/Raman microscopy	Aqueous	C-based	C-based ENMs	DOC	100	10 nm[e]	No	AFM, SEM
Pyrolysis/TGA-GC-MS	Solid	C-based	C-based ENMs	DOC	100	NA[d]	No	TGA, FTIR
spICPMS	All[f]	Inorganic	heavy elements	Particles having the same elements in the same ratio, adsorbed metals on particle surfaces	1	5	Yes	LA, sp-ICP-TOF-MS, FFF
NIR-fluorescence	Aqueous (solid w/microscopy)	C-based	SWCNTs	NIR-absorbing molecules	1	NA[d]	No	
NTA	Aqueous	Both	Highly scattering ENMs	Large particles that scatter highly	100	20	Yes	Fluorescence

Note that this is a summary only, and that all properties come with certain caveats that are explained more thoroughly in the text.
[a] The sample types that can be measured without extensive sample pretreatment such as extraction.
[b] The LOD refers to the lowest possible LOD that can be attained, usually only for one "ideal" unaggregated material using the best instrument available.
[c] When coupled to Raman and/or infrared measurements.
[d] This technique cannot quantify sizes.
[e] This LOD is when IR/Raman is used in combination with AFM.
[f] Solid samples only when coupled to laser ablation (LA).

References

Arakawa, A., et al., 2019. High-resolution laser ablation inductively coupled plasma mass spectrometry used to study transport of metallic nanoparticles through collagen-rich microstructures in fibroblast multicellular spheroids. Anal. Bioanal. Chem. 411, 3497–3506.

Arvidsson, R., Molander, S., Sandén, B.A., 2012. Assessing the environmental risks of silver from clothes in an urban area. Hum. Ecol. Risk Assess. Int. J. 20 (4), 1008–1022.

Arvidsson, R., et al., 2018. "just carbon": ideas about graphene risks by graphene researchers and innovation advisors. NanoEthics 12 (3), 199–210.

Baalousha, M., et al., 2006. Size fractionation and characterization of natural colloids by flow-field flow fractionation coupled to multi-angle laser light scattering. J. Chromatogr. A 1104 (1–2), 272–281.

Becker, J.S., et al., 2007. Quantitative imaging of selenium, copper, and zinc in thin sections of biological tissues (slugs-genus Arion) measured by laser ablation inductively coupled plasma mass spectrometry. Anal. Chem. 79 (16), 6074–6080.

Beckett, R., et al., 1988. Separation and size characterization of colloidal particles in river water by sedimentation field-flow fractionation. Water Res. 22 (12), 1535–1545.

Benesova, I., et al., 2016. Direct analysis of gold nanoparticles from dried droplets using substrate-assisted laser desorption single particle-ICPMS. Anal. Chem. 88 (5), 2576–2582.

Benettoni, P., et al., 2019. Identification of nanoparticles and their localization in algal biofilm by 3D-imaging secondary ion mass spectrometry. J. Anal. At. Spectrom. 34 (6), 1098–1108.

Benoit, G., et al., 1994. Partitioning of Cu, Pb, Ag, Zn, Fe, Al, and Mn between filter-retained particles, colloids, and solution in six Texas estuaries. Mar. Chem. 45 (4), 307–336.

Bisesi Jr., J.H., et al., 2014. Tracking and quantification of single-walled carbon nanotubes in fish using near- infrared fluorescence. Environ. Sci. Technol. 48 (3), 1973–1983.

Bläsing, M., Amelung, W., 2018. Plastics in soil: analytical methods and possible sources. Sci. Total Environ. 612, 422–435.

Bohme, S., et al., 2015. Exploring LA-ICP-MS as a quantitative imaging technique to study nanoparticle uptake in Daphnia magna and zebrafish (Danio rerio) embryos. Anal. Bioanal. Chem. 407 (18), 5477–5485.

Bone, A.J., et al., 2012. Biotic and abiotic interactions in aquatic microcosms determine fate and toxicity of Ag nanoparticles: part 2-toxicity and Ag speciation. Environ. Sci. Technol. 46 (13), 6925–6933.

Bootz, A., et al., 2004. Comparison of scanning electron microscopy, dynamic light scattering and analytical ultracentrifugation for the sizing of poly(butyl cyanoacrylate) nanoparticles. Eur. J. Pharm. Biopharm. 57 (2), 369–375.

Borovinskaya, O., et al., 2013. A prototype of a new inductively coupled plasma time-of-flight mass spectrometer providing temporally resolved, multi-element detection of short signals generated by single particles and droplets. J. Anal. At. Spectrom. 28 (2), 226–233.

Borovinskaya, O., et al., 2014. Simultaneous mass quantification of nanoparticles of different composition in a mixture by microdroplet generator-ICPTOFMS. Anal. Chem. 86 (16), 8142–8148.

Buesseler, K.O., et al., 1996. An intercomparison of cross-flow filtration techniques used for sampling marine colloids: overview and organic carbon results. Mar. Chem. 55 (1), 1–31.

Bundschuh, M., et al., 2018. Nanoparticles in the environment: where do we come from, where do we go to? Environ. Sci. Eur. 30 (1), 6.

Cornelis, G., et al., 2010. A method for determining the partitioning of manufactured silver and cerium oxide nanoparticles in soil environments. Environ. Chem. 7 (3), 298–308.

Cornelis, G., et al., 2014. Fate and bioavailability of engineered nanoparticles in soils: a review. Crit. Rev. Environ. Sci. Technol. 44, 2720–2764.

Dazzi, A., Prater, C.B., 2017. AFM-IR: technology and applications in nanoscale infrared spectroscopy and chemical imaging. Chem. Rev. 117 (7), 5146–5173.

Dazzi, A., et al., 2012. AFM-IR: combining atomic force microscopy and infrared spectroscopy for nanoscale chemical characterization. Appl. Spectrosc. 66 (12), 1365–1384.

De Leeuw, J.W., et al., 1986. Screening of anthropogenic compounds in polluted sediments and soils by flash evaporation/pyrolysis gas chromatography-mass spectrometry. Anal. Chem. 58 (8), 1852–1857.

Degueldre, C., Favarger, P.Y., 2003. Colloid analysis by single particle inductively coupled plasma-mass spectroscopy: a feasibility study. Colloids Surf. A Physicochem. Eng. Asp. 217 (1–3), 137–142.

Degueldre, C., Favarger, P.Y., Wold, S., 2006. Gold colloid analysis by inductively coupled plasma-mass spectrometry in a single particle mode. Anal. Chim. Acta 555 (2), 263–268.

Dekiff, J.H., et al., 2014. Occurrence and spatial distribution of microplastics in sediments from Norderney. Environ. Pollut. 186, 248–256.

Domingos, R.F., et al., 2009. Characterizing manufactured nanoparticles in the environment: multimethod determination of particle sizes. Environ. Sci. Technol. 43 (19), 7277–7284.

Donaldson, K., et al., 2013. The biologically effective dose in inhalation nanotoxicology. Acc. Chem. Res. 46 (3), 723–732.

Doolette, C.L., et al., 2013. Transformation of PVP coated silver nanoparticles in a simulated wastewater treatment process and the effect on microbial communities. Chem. Cent. J. 7 (1), 46.

Dümichen, E., et al., 2015. Analysis of polyethylene microplastics in environmental samples, using a thermal decomposition method. Water Res. 85, 451–457.

Dümichen, E., et al., 2017. Fast identification of microplastics in complex environmental samples by a thermal degradation method. Chemosphere 174, 572–584.

Dziewatkoski, M.P., Daniels, L.B., Olesik, J.W., 1996. Time-resolved inductively coupled plasma mass spectrometry measurements with individual, monodisperse drop sample introduction. Anal. Chem. 68 (7), 1101–1109.

European Commission, 2011. Commission recommendation of 18 October 2011 on the definition of nanomaterial. In: Official Journal of the European Union. T.E. commission, Editor, Brussels.

Fabbri, D., Tartari, D., Trombini, C., 2000. Analysis of poly(vinyl chloride) and other polymers in sediments and suspended matter of a coastal lagoon by pyrolysis-gas chromatography-mass spectrometry. Anal. Chim. Acta 413 (1), 3–11.

Farré, M., Sanchís, J., Barceló, D., 2011. Analysis and assessment of the occurrence, the fate and the behavior of nanomaterials in the environment. TrAC Trends Anal. Chem. 30 (3), 517–527.

Fischer, M., Scholz-Böttcher, B.M., 2017. Simultaneous trace identification and quantification of common types of microplastics in environmental samples by pyrolysis-gas chromatography–mass spectrometry. Environ. Sci. Technol. 51 (9), 5052–5060.

Franklin, N.M., et al., 2007. Comparative toxicity of nanoparticulate ZnO, bulk ZnO, and $ZnCl_2$ to a freshwater microalga (Pseudokirchneriella subcapitata): the importance of particle solubility. Environ. Sci. Technol. 41 (24), 8484–8490.

Fries, E., et al., 2013. Identification of polymer types and additives in marine microplastic particles using pyrolysis-GC/MS and scanning electron microscopy. Environ Sci Process Impacts 15 (10), 1949–1956.

Gallego-Urrea, J.A., et al., 2010. Measurements of nanoparticle number concentrations and size distributions in contrasting aquatic environments using nanoparticle tracking analysis. Environ. Chem. 7 (1), 67–81.

Gallego-Urrea, J.A., Tuoriniemi, J., Hassellöv, M., 2011. Applications of particle-tracking analysis to the determination of size distributions and concentrations of nanoparticles in environmental, biological and food samples. TrAC, Trends Anal. Chem. 30 (3), 473–483.

Gallego-Urrea, J.A., et al., 2014. Multimethod 3D characterization of natural plate-like nanoparticles: shape effects on equivalent size measurements. J. Nanopart. Res. 16 (5).

Gallego-Urrea, J.A., et al., 2016. Coagulation and sedimentation of gold nanoparticles and illite in model natural waters: influence of initial particle concentration. NanoImpact 3-4, 67–74.

Giddings, J.C., 1993. Field-flow fractionation - analysis of macromolecular, colloidal, and particulate materials. Science 260 (5113), 1456–1465.

Giddings, J.C., Yang, F.J., Myers, M.N., 1977. Flow field-flow fractionation as a methodology for protein separation and characterization. Anal. Biochem. 81 (2), 395–407.

Gigault, J., et al., 2014. Rational strategy for characterization of nanoscale particles by asymmetric-flow field flow fractionation: a tutorial. Anal. Chim. Acta 809, 9–24.

Gigault, J., et al., 2018. Current opinion: what is a nanoplastic? Environ. Pollut. 235, 1030–1034.

Gimbert, L.J., et al., 2005. Comparison of centrifugation and filtration techniques for the size fractionation of colloidal material in soil suspensions using sedimentation field-flow fractionation. Environ. Sci. Technol. 39 (6), 1731–1735.

Gondikas, A.P., Jang, E.K., Hsu-Kim, H., 2010. Influence of amino acids cysteine and serine on aggregation kinetics of zinc and mercury sulfide colloids. J. Colloid Interface Sci. 347 (2), 167–171.

Gondikas, A.P., et al., 2012. Cysteine-induced modifications of zero-valent silver nanomaterials: implications for particle surface chemistry, aggregation, dissolution, and silver speciation. Environ. Sci. Technol. 120406095922008.

Gondikas, A.P., et al., 2014. Release of TiO2 nanoparticles from sunscreens into surface waters: a one-year survey at the old Danube recreational lake. Environ. Sci. Technol. 48 (10), 5415–5422.

Gondikas, A., et al., 2018. Where is the nano? Analytical approaches for the detection and quantification of TiO2 engineered nanoparticles in surface waters. Environ. Sci.: Nano 5 (2), 313–326.

Gottschalk, F., Nowack, B., 2011. The release of engineered nanomaterials to the environment. J. Environ. Monit. 13 (5), 1145–1155.

Gottschalk, F., et al., 2009. Modeled environmental concentrations of engineered nanomaterials (TiO_2, ZnO, Ag, CNT, fullerenes) for different regions. Environ. Sci. Technol. 43 (24), 9216–9222.

Gottschalk, F., Sun, T., Nowack, B., 2013. Environmental concentrations of engineered nanomaterials: review of modeling and analytical studies. Environ. Pollut. 181, 287–300.

Gundlach-Graham, A., et al., 2018. High-resolution, quantitative element imaging of an upper crust, low-angle Cataclasite (Zuccale Fault, Northern Apennines) by laser ablation ICP time-of-flight mass spectrometry. Geostand. Geoanal. Res. 42 (4), 559–574.

Gupta, S., Aten, C., 1993. Comparison and evaluation of extraction media and their suitability in a simple model to predict the biological relevance of heavy metal concentrations in contaminated soils. Int. J. Environ. Anal. Chem. 51 (1–4), 25–46.

Hagendorfer, H., et al., 2011. Application of an asymmetric flow field flow fractionation multi-detector approach for metallic engineered nanoparticle characterization—prospects and limitations demonstrated on Au nanoparticles. Anal. Chim. Acta 706 (2), 367–378.

Handy, R.D., Shaw, B.J., 2007. Ecotoxicity of nanomaterials to fish: challenges for ecotoxicity testing. Integr. Environ. Assess. Manag. 3 (3), 458–460.

Handy, R.D., et al., 2012. Ecotoxicity test methods for engineered nanomaterials: practical experiences and recommendations from the bench. Environ. Toxicol. Chem. 31 (1), 15–31.

Hassellöv, M., Kaegi, R., 2010. Analysis and characterization of manufactured nanoparticles in aquatic environments. In: Analysis. Wiley Online Library, pp. 211–266.

Hassellöv, M., et al., 1999. Determination of continuous size and trace element distribution of colloidal material in natural water by on-line coupling of flow field-flow fractionation with ICPMS. Anal. Chem. 71 (16), 3497–3502.

Hassellöv, M., et al., 2008. Nanoparticle analysis and characterization methodologies in environmental risk assessment of engineered nanoparticles. Ecotoxicology 17 (5), 344–361.

Hendriks, L., et al., 2017. Characterization of a new ICP-TOFMS instrument with continuous and discrete introduction of solutions. J. Anal. At. Spectrom. 32 (3), 548–561.

Hendriks, L., Gundlach-Graham, A., Günther, D., 2018. Analysis of inorganic nanoparticles by single-particle inductively coupled plasma time-of-flight mass spectrometry. CHIMIA Int. J. Chem. 72 (4), 221–226.

Hernandez, L.M., Yousefi, N., Tufenkji, N., 2017. Are there nanoplastics in your personal care products? Environ. Sci. Technol. Lett. 4 (7), 280–285.

Herrero-Latorre, C., et al., 2015. Characterization of carbon nanotubes and analytical methods for their determination in environmental and biological samples: a review. Anal. Chim. Acta 853, 77–94.

Hidalgo-Ruz, V., et al., 2012. Microplastics in the marine environment: a review of the methods used for identification and quantification. Environ. Sci. Technol. 46 (6), 3060–3075.

Horton, A.A., et al., 2017. Microplastics in freshwater and terrestrial environments: evaluating the current understanding to identify the knowledge gaps and future research priorities. Sci. Total Environ. 586, 127–141.

Houk, R.S., et al., 1980. Inductively coupled argon plasma as an ion source for mass spectrometric determination of trace elements. Anal. Chem. 52 (14), 2283–2289.

Huynh, K.A., et al., 2016. Detection and quantification of silver nanoparticles at environmentally relevant concentrations using asymmetric flow field–flow fractionation online with single particle inductively coupled plasma mass spectrometry. Anal. Chem. 88 (9), 4909–4916.

Imhof, H.K., et al., 2016. Pigments and plastic in limnetic ecosystems: a qualitative and quantitative study on microparticles of different size classes. Water Res. 98, 64–74.

International Organization for Standardization, 2009. Application of Inductively Coupled Plasma Mass Spectrometry (ICP-MS) - Part 2: Determination of 62 Elements. ISO, Geneva, Switzerland.

International Organization for Standardization, 2018. Nanotechnologies—Analysis of Nano-Objects Using Asymmetrical-Flow and Centrifugal Field-Flow Fractionation. ISO.

Kaegi, R., et al., 2011. Behavior of metallic silver nanoparticles in a pilot wastewater treatment plant. Environ. Sci. Technol. 45.

Kaegi, R., et al., 2013. Fate and transformation of silver nanoparticles in urban wastewater systems. Water Res. 47 (12), 3866–3877.

Kappler, A., et al., 2016. Analysis of environmental microplastics by vibrational microspectroscopy: FTIR, Raman or both? Anal. Bioanal. Chem. 408 (29), 8377–8391.

Kaszuba, M., et al., 2008. Measuring sub nanometre sizes using dynamic light scattering. J. Nanopart. Res. 10 (5), 823–829.

Kim, P., et al., 1999. Electronic density of states of atomically resolved single-walled carbon nanotubes: Van Hove singularities and end states. Phys. Rev. Lett. 82 (6), 1225.

Klaine, S.J., et al., 2008. Nanomaterials in the environment: behavior, fate, bioavailability, and effects. Environ. Toxicol. Chem. 27 (9), 1825–1851.

Laborda, F., et al., 2011. Selective identification, characterization and determination of dissolved silver(i) and silver nanoparticles based on single particle detection by inductively coupled plasma mass spectrometry. J. Anal. At. Spectrom. 26 (7).

Laborda, F., et al., 2013a. Critical considerations for the determination of nanoparticle number concentrations, size and number size distributions by single particle ICP-MS. J. Anal. At. Spectrom. 28 (8), 1220–1232.

Laborda, F., Bolea, E., Jimenez-Lamana, J., 2013b. Single particle inductively coupled plasma mass spectrometry: a powerful tool for nanoanalysis. Anal. Chem. 86 (5), 2270–2278.

Laborda, F., et al., 2016. Detection, characterization and quantification of inorganic engineered nanomaterials: a review of techniques and methodological approaches for the analysis of complex samples. Anal. Chim. Acta 904.

Lee, W.-W., Chan, W.-T., 2015. Calibration of single-particle inductively coupled plasma-mass spectrometry (SP-ICP-MS). J. Anal. At. Spectrom. 30 (6), 1245–1254.

Lee, S., et al., 2014. Nanoparticle size detection limits by single particle ICP-MS for 40 elements. Environ. Sci. Technol. 48 (17), 10291–10300.

Lenz, R., et al., 2015. A critical assessment of visual identification of marine microplastic using Raman spectroscopy for analysis improvement. Mar. Pollut. Bull. 100 (1), 82–91.

Levard, C., et al., 2012. Environmental transformations of silver nanoparticles: impact on stability and toxicity. Environ. Sci. Technol. 120229163713001.

Linders, T., et al., 2018. Particle sources and transport in stratified Nordic coastal seas in the Anthropocene. Elementa 6.

Liu, J., Hurt, R.H., 2010. Ion release kinetics and particle persistence in aqueous nano-silver colloids. Environ. Sci. Technol. 44 (6), 2169–2175.

Liu, J., et al., 2014. Capabilities of single particle inductively coupled plasma mass spectrometry for the size measurement of nanoparticles: a case study on gold nanoparticles. Anal. Chem. 86 (7), 3405–3414.

Liu, J., et al., 2017. Overcoming challenges in single particle inductively coupled plasma mass spectrometry measurement of silver nanoparticles. Anal. Bioanal. Chem. 409 (25), 6027–6039.

Löder, M.G.J., et al., 2015. Focal plane array detector-based micro-Fourier-transform infrared imaging for the analysis of microplastics in environmental samples. Environ. Chem. 12 (5), 563–581.

Loeschner, K., et al., 2013. Optimization and evaluation of asymmetric flow field-flow fractionation of silver nanoparticles. J. Chromatogr. A 1272, 116–125.

Loosli, F., et al., 2019a. Sewage spills are a major source of titanium dioxide engineered (nano)-particle release into the environment. Environ. Sci.: Nano 6 (3), 763–777.

Loosli, F., et al., 2019b. Improved extraction efficiency of natural nanomaterials in soils to facilitate their characterization using a multimethod approach. Sci. Total Environ. 677, 34–46.

Ma, R., et al., 2014. Fate of zinc oxide and silver nanoparticles in a pilot wastewater treatment plant and in processed biosolids. Environ. Sci. Technol. 48 (1), 104–112.

Malloy, A., Carr, B., 2006. NanoParticle tracking analysis - the halo™ system. Part. Part. Syst. Charact. 23 (2), 197–204.

Mavrocordatos, D., Perret, D., Leppard, G.G., 2007. Strategies and advances in the characterisation of environmental colloids by electron microscopy. In: Wilkinson, K.J., Lead, J.R. (Eds.), Environmental Colloids and Particles. Behaviour, Separation and Characterisation. John Wiley & Sons, Chichester, UK.

Meesters, J.A.J., et al., 2014. Multimedia modeling of engineered nanoparticles with SimpleBox4nano: model definition and evaluation. Environ. Sci. Technol. 48 (10), 5726–5736.

Meesters, J.A.J., et al., 2016. Multimedia environmental fate and speciation of engineered nanoparticles: a probabilistic modeling approach. Environ. Sci.: Nano 3 (4), 715–727.

Mehrabi, K., et al., 2017. Improvements in nanoparticle tracking analysis to measure particle aggregation and mass distribution: a case study on engineered nanomaterial stability in incineration landfill leachates. Environ. Sci. Technol. 51 (10), 5611–5621.

Meier, C., et al., 2016. Transformation of silver nanoparticles in sewage sludge during incineration. Environ. Sci. Technol. 50 (7), 3503–3510.

Meisterjahn, B., et al., 2016. Silver and gold nanoparticle separation using asymmetrical flow-field flow fractionation: influence of run conditions and of particle and membrane charges. J. Chromatogr. A 1440, 150–159.

Meyns, M., Finckh, S., Gerdts, G., 2018. nano-FTIR - from defined to environmental (nano) plastic samples. In: MICRO 2018, Lanzarote.

Michen, B., et al., 2015. Avoiding drying-artifacts in transmission electron microscopy: characterizing the size and colloidal state of nanoparticles. Sci. Rep. 5, 9793.

Montaño, M.D., et al., 2014. Improvements in the detection and characterization of engineered nanoparticles using spICP-MS with microsecond dwell times. Environ. Sci.: Nano 1 (4), 338–346.

Montano, M., et al., 2014. Current status and future direction for examining engineered nanoparticles in natural systems. Environ. Chem. 11 (4), 351–366.

Montaño, M.D., et al., 2016a. Single particle ICP-MS: advances toward routine analysis of nanomaterials. Anal. Bioanal. Chem. 408 (19), 5053–5074.

Montaño, M.D., et al., 2016b. Methods for the detection and characterization of silica colloids by microsecond spICP-MS. Anal. Chem. 88 (9), 4733–4741.

Myers, D., et al., 1994. An inductively coupled plasma—time-of-flight mass spectrometer for elemental analysis. Part I: optimization and characteristics. J. Am. Soc. Mass Spectrom. 5 (11), 1008–1016.

Naasz, S., et al., 2018. Multi-element analysis of single nanoparticles by ICP-MS using quadrupole and time-of-flight technologies. J. Anal. At. Spectrom. 33 (5), 835–845.

Narkis, N., Rebhun, M., 1975. The mechanism of flocculation processes in the presence of humic substances. J. - Am. Water Works Assoc. 67 (2), 101–108.

Nowack, B., 2017. Evaluation of environmental exposure models for engineered nanomaterials in a regulatory context. NanoImpact 8, 38–47.

Nowack, B., et al., 2012. Potential scenarios for nanomaterial release and subsequent alteration in the environment. Environ. Toxicol. Chem. 31 (1), 50–59.

Nowack, B., et al., 2015. Progress towards the validation of modeled environmental concentrations of engineered nanomaterials by analytical measurements. Environ. Sci.: Nano 2 (5), 421–428.

O'connell, M.J., et al., 2002. Band gap fluorescence from individual single-walled carbon nanotubes. Science 297 (5581), 593–596.

Österreicher, J.A., et al., 2018. Information depth in backscattered electron microscopy of nanoparticles within a solid matrix. Mater. Charact. 138, 145–153.

Pace, H.E., et al., 2012. Single particle inductively coupled plasma-mass spectrometry: a performance evaluation and method comparison in the determination of nanoparticle size. Environ. Sci. Technol. 46, 12272–12280.

Park, H.-J., et al., 2013. Removal characteristics of engineered nanoparticles by activated sludge. Chemosphere 92 (5), 524–528.

Petersen, E.J., et al., 2014. Identification and avoidance of potential artifacts and misinterpretations in nanomaterial ecotoxicity measurements. Environ. Sci. Technol. 48 (8), 4226–4246.

Petersen, E.J., et al., 2016. Quantification of carbon nanotubes in environmental matrices: current capabilities, case studies, and future prospects. Environ. Sci. Technol. 50 (9), 4587–4605.

Petosa, A.R., et al., 2010. Aggregation and deposition of engineered nanomaterials in aquatic environments: role of physicochemical interactions. Environ. Sci. Technol. 44 (17), 6532–6549.

Philippe, A., et al., 2018. Characterization of the natural colloidal TiO2 background in soil. Separations 5 (4), 50.

Piccinno, F., et al., 2012. Industrial production quantities and uses of ten engineered nanomaterials in Europe and the world. J. Nanopart. Res. 14 (9), 1109.

Pirrie, D., et al., 2009. Automated SEM-EDS (QEMSCAN®). Mineral analysis in forensic soil investigations: testing instrumental reproducibility. In: Ritz, K., Dawson, L., Miller, D. (Eds.), Criminal and Environmental Soil Forensics. Springer.

Plathe, K., et al., 2010. Using FlFFF and aTEM to determine trace metal–nanoparticle associations in riverbed sediment. Environ. Chem. 7 (1), 82–93.

Plathe, K.L., et al., 2013. The role of nanominerals and mineral nanoparticles in the transport of toxic trace metals: field-flow fractionation and analytical TEM analyses after nanoparticle isolation and density separation. Geochim. Cosmochim. Acta 102, 213–225.

Praetorius, A., Scheringer, M., Hungerbühler, K., 2012. Development of environmental fate models for engineered nanoparticles—a case study of TiO2 nanoparticles in the Rhine River. Environ. Sci. Technol. 46 (12), 6705–6713.

Praetorius, A., et al., 2017. Single-particle multi-element fingerprinting (spMEF) using inductively coupled plasma time of flight mass spectrometry (ICP-TOFMS) to identify engineered nanoparticles against the elevated natural background in soils. Environ. Sci.: Nano 4 (2), 307–314.

Rakcheev, D., Philippe, A., Schaumann, G.E., 2013. Hydrodynamic chromatography coupled with single particle-inductively coupled plasma mass spectrometry for investigating nanoparticles agglomerates. Anal. Chem. 85 (22), 10643–10647.

Reed, R.B., et al., 2017. Multi-day diurnal measurements of Ti-containing nanoparticle and organic sunscreen chemical release during recreational use of a natural surface water. Environ. Sci.: Nano 4 (1), 69–77.

Röding, M., et al., 2011. Measuring absolute number concentrations of nanoparticles using single-particle tracking. Phys. Rev. E 84 (3), 031920.

Sanchís, J., et al., 2015. Nanoparticle tracking analysis characterisation and parts-per-quadrillion determination of fullerenes in river samples from Barcelona catchment area. Anal. Bioanal. Chem. 407 (15), 4261–4275.

Saveyn, H., et al., 2010. Accurate particle size distribution determination by nanoparticle tracking analysis based on 2-D Brownian dynamics simulation. J. Colloid Interface Sci. 352, 593–600.

Schierz, A., et al., 2012. Characterization and quantitative analysis of single-walled carbon nanotubes in the aquatic environment using near-infrared fluorescence spectroscopy. Environ. Sci. Technol. 46 (22), 12262–12271.

Schierz, A., et al., 2014. Fate of single-walled carbon nanotubes in wetland ecosystems. Environ. Sci.: Nano 1 (6), 574–583.

Schmid, O., Stoeger, T., 2016. Surface area is the biologically most effective dose metric for acute nanoparticle toxicity in the lung. J. Aerosol Sci. 99, 133–143.

Schmid, T., et al., 2013. Nanoscale chemical imaging using tip-enhanced Raman spectroscopy: a critical review. Angew. Chem. Int. Ed. 52 (23), 5940–5954.

Schwertfeger, D.M., et al., 2017. Extracting metallic nanoparticles from soils for quantitative analysis: method development using engineered silver nanoparticles and SP-ICP-MS. Anal. Chem. 89 (4), 2505–2513.

Simon, M., van Alst, N., Vollertsen, J., 2018. Quantification of microplastic mass and removal rates at wastewater treatment plants applying focal plane Array (FPA)-based Fourier transform infrared (FT-IR) imaging. Water Res. 142, 1–9.

Skjolding, L.M., et al., 2016. Aquatic ecotoxicity testing of nanoparticles-the quest to disclose nanoparticle effects. Angew. Chem. Int. Ed. Engl. 55 (49), 15224–15239.

Song, Y.K., et al., 2015. A comparison of microscopic and spectroscopic identification methods for analysis of microplastics in environmental samples. Mar. Pollut. Bull. 93 (1), 202–209.

Stark, H.J., Wennrich, R., 2011. A new approach for calibration of laser ablation inductively coupled plasma mass spectrometry using thin layers of spiked agarose gels as references. Anal. Bioanal. Chem. 399 (6), 2211–2217.

Stolpe, B., Hassellöv, M., 2007. Changes in size distribution of fresh water nanoscale colloidal matter and associated elements on mixing with seawater. Geochim. Cosmochim. Acta 71 (13), 3292–3301.

Sun, T.Y., et al., 2014. Comprehensive probabilistic modelling of environmental emissions of engineered nanomaterials. Environ. Pollut. 185, 69–76.

Sun, T.Y., et al., 2016. Dynamic probabilistic modelling of environmental emissions of engineered nanomaterials. Environ. Sci. Technol. 50, 4701–4711.

Tiede, K., et al., 2010. Application of hydrodynamic chromatography-ICP-MS to investigate the fate of silver nanoparticles in activated sludge. J. Anal. At. Spectrom. 25 (7), 1149–1154.

Toncelli, C., et al., 2016. Flow injection with on-line dilution and single particle inductively coupled plasma – mass spectrometry for monitoring silver nanoparticles in seawater and in marine microorganisms. J. Anal. At. Spectrom. 31 (7), 1430–1439.

Tuoriniemi, J., Cornelis, G., Hassellöv, M., 2012. Size discrimination and detection capabilities of single-particle ICP-MS for environmental analysis of silver nanoparticles. Anal. Chem. 29, 743–752.

Tuoriniemi, J., Cornelis, G., Hassellöv, M., 2014a. Improving accuracy of single particle ICPMS for measurement of size distributions and number concentrations of nanoparticles by determining analyte partitioning during nebulisation. J. Anal. At. Spectrom. 29, 743–752.

Tuoriniemi, J., et al., 2014b. Intermethod comparison of the particle size distributions of colloidal silica nanoparticles. Sci. Technol. Adv. Mater. 15 (3), 035009.

Tuoriniemi, J., et al., 2014c. In situ characterisation of physicochemical state and concentration of nanoparticles in soil ecotoxicity studies using environmental scanning electron microscopy. Environ. Chem. 11 (4), 367–376.

Tuoriniemi, J., et al., 2017. Size dependence of silver nanoparticle removal in a wastewater treatment plant mesocosm measured by FAST single particle ICP-MS. Environ. Sci.: Nano 4 (5), 1189–1197.

Utembe, W., et al., 2015. Dissolution and biodurability: important parameters needed for risk assessment of nanomaterials. Part. Fibre Toxicol. 12, 11.

Van Hoecke, K., et al., 2008. Ecotoxicity of silica nanoparticles to the green alga Pseudokirchneriella subcapitata: importance of surface area. Environ. Toxicol. Chem. 27 (9), 1948–1957.

Van Hoecke, K., et al., 2011. Aggregation and ecotoxicity of CeO2 nanoparticles in synthetic and natural waters with variable pH, organic matter concentration and ionic strength. Environ. Pollut. 159 (4), 970–976.

Velzeboer, I., et al., 2008. Nanomaterials in the environment aquatic ecotoxicity tests of some nanomaterials. Environ. Toxicol. Chem. 27 (9), 1942–1947.

von der Kammer, F., Baborowski, M., Friese, K., 2005. Field-flow fractionation coupled to multi-angle laser light scattering detectors: applicability and analytical benefits for the analysis of environmental colloids. Anal. Chim. Acta 552 (1), 166–174.

Von der Kammer, F., et al., 2012. Analysis of engineered nanomaterials in complex matrices (environment and biota): general considerations and conceptual case studies. Environ. Toxicol. Chem. 31 (1), 32–49.

Wagner, S., et al., 2014. Spot the difference: engineered and natural nanoparticles in the environment—release, behavior, and fate. Angew. Chem. Int. Ed. 53 (46), 12398–12419.

Wagner, S., et al., 2015. First steps towards a generic sample preparation scheme for inorganic engineered nanoparticles in a complex matrix for detection, characterization, and quantification by asymmetric flow-field flow fractionation coupled to multi-angle light scattering and ICP-MS. J. Anal. At. Spectrom. 30 (6), 1286–1296.

Wahlund, K.-G., 2013. Flow field-flow fractionation: critical overview. J. Chromatogr. A 1287, 97–112.

Wang, Y., Nowack, B., 2018. Environmental risk assessment of engineered nano-SiO2, nano iron oxides, nano-CeO2, nano-Al2O3, and quantum dots. Environ. Toxicol. Chem. 37 (5), 1387–1395.

Welsher, K., et al., 2008. Selective probing and imaging of cells with single walled carbon nanotubes as near-infrared fluorescent molecules. Nano Lett. 8 (2), 586–590.

Williams, D.B., Carter, C.B., 1996. The transmission electron microscope. In: Transmission Electron Microscopy. Springer, pp. 3–17.

Winterwerp, J.C., 1998. A simple model for turbulence induced flocculation of cohesive sediment. J. Hydraul. Res. 36 (3), 309–326.

WITec GmbH, 2018. On the RISE: Correlative Confocal Raman and Scanning Electron Microscopy. cited 2019; Available from: https://www.witec.de/assets/Literature/Files/WITec-AppNote-RISE-Raman-SEM-web.pdf.

Zanker, H., Schierz, A., 2012. Engineered nanoparticles and their identification among natural nanoparticles. Annu Rev Anal Chem (Palo Alto, Calif) 5, 107–132.

Zetsche, E.-M., Ploug, H., 2015. Marine chemistry special issue: particles in aquatic environments: from invisible exopolymers to sinking aggregates. Mar. Chem. 175 (Supplement C), 1–4.

Zhao, S., et al., 2018. Limitations for microplastic quantification in the ocean and recommendations for improvement and standardization. In: Zeng, E.Y. (Ed.), Microplastic Contamination in Aquatic Environments. Elsevier, pp. 27–49 (Chapter 2).

Citizen science monitoring of air pollution: Challenges and experiences from the six iSCAPE living labs

5

Francesco Pilla[a], Katinka Schaaf[b], and Lee Markham[b]
[a]Spatial Dynamics Lab, College of Engineering, University College Dublin, Dublin, Ireland,
[b]Connected Places Catapult, London, United Kingdom

1 Introduction

This chapter presents the citizen science (CS) activities aimed at air pollution monitoring with low-cost sensors (the Smart Citizen Kits[a]) undertaken by the six iSCAPE Living Labs (LLs) with a variety of local stakeholders. iSCAPE is a H2020 project funded by the European Commission and aims at improving the smart control of air pollution in European cities.[b] The iSCAPE LLs are available in Bologna, Bottrop, Dublin, Guildford, Hasselt, and Vantaa. The LLs followed a CS framework developed by the partner in charge of this task, Connected Places Catapult (CPC), which resulted in the participation of approximately 100 participants across the six LL cities. The goal of the workshops was to engage with the public as well as to establish a CS community.

The short-term goal of the workshops was to educate the LLs and the general public about CS, use and test the Smart Citizen Kits, and build an air quality monitoring community motivated to influence and change local policies around air pollution. The long-term objective was to have citizen scientists actively working and collaborating with the LLs, from working on their self-initiated projects to assisting the LLs.

The iSCAPE Smart Citizen Kit (SCK) is a low-cost urban air quality sensor and is at the core of the framework. The first version of the sensing kit was crowdfunded by a kick-starter campaign led by the Fab Lab Barcelona in 2013. The second version (SCK 2.0) was made available to citizens and the LLs for the iSCAPE project in 2018. Its use during the iSCAPE project has directly contributed to SCK 2.1, which is now commercially available.

The SCK is the core of what we call the Smart Citizen System—a complete set of modular hardware components aiming to provide tools for environmental monitoring, ranging from citizen science and educational activities to more advanced scientific

[a] https://smartcitizen.me/kits/.

[b] https://www.iscapeproject.eu/.

Monitoring Environmental Contaminants. https://doi.org/10.1016/B978-0-444-64335-3.00006-2

research. The system is designed in an extendable way, with a central data logger (the Data Board) with network connectivity from which the different components are branched. The system is based on the principle of reproducibility, also integrating non-hardware components such as a dedicated storage platform and a sensor analysis framework (Camprodon et al., 2019).

The SCK 2.1 sensors are listed in the following table:

Measurement	Units	Sensors
Air temperature	°C	Sensirion SHT-31
Relative humidity	% REL	Sensirion SHT-31
Noise level	dBA	InvenSense ICS-434342
Ambient light	Lux	Rohm BH1721FVC
Barometric pressure	Pa	NXP MPL3115A26
Equivalent carbon dioxide	ppm	AMS CCS811
Volatile organic compounds	ppb	AMS CCS811
Particulate matter PM 1/2.5/10	$\mu g/m^3$	Plantower PMS 5003

2 Methodology: The iSCAPE citizen science framework

The workshop ran in the six iSCAPE Living Labs aimed at engaging citizens in air pollution control in cities by providing them with a scientific task, monitoring exposure to air pollutants, and the tools to carry it out, the low-cost sensing kits. All the activities were designed following a citizen science framework.

There is no universally accepted definition of CS, but all the definitions agree on the main focus, which is the general public being actively engaged in scientific research activities.

The first step in delivering a total of 12 workshops (two CS workshops in each of the LL cities) was to create an iSCAPE-specific CS framework, which would ensure consistency across the cities and could be easily configurable by nonexperts/aspiring facilitators. The framework included a very comprehensive set of guidelines and tools, fully designed worksheets for both sessions, and predesigned PowerPoint slides and agendas.

Overall, the iSCAPE framework is made up of seven steps and can be delivered with a time frame of 2–3 months. The CS approach can easily extend beyond the two CS workshops undertaken by each LL, as the process can be repeated and iterated based on the learning and engagement of the citizens taking part.

The steps of the framework are as follows:

1. Preparation and promotion of the CS workshops
2. Delivering the first workshop—Onboarding, planning, and sensing part 1
3. Data collection (support of the teams during the data collection phase)
4. Data analysis and visualization
5. Delivering the second workshop—Sensing part 2, awareness, action, and reflection

2.1 Preparation and promotion of the CS workshops

Delivering the two CS workshops in a short time frame and with limited resources was a major challenge for the LLs. The crucial question that the LLs needed to answer was how to reach out to and recruit a suitable audience that is both interested and motivated to engage with the LL for a minimum of 1 month (two workshops)?

Below are the three golden workshop rules (taken directly from the CS Pack):

1. *Easy to attend*: The workshop should be held at a place that is easy to reach and at a time that is suitable to increase the number and variety of participants.
2. *Fun and entertaining*: The workshop should show the participants something they did not know or teach them a new skill.
3. *Memorable so people come back*: If the events stay in people's mind, they will talk about them to others, which promotes iSCAPE and the individual LL! This also means they will return to the second or even third round of the CS workshops.

2.2 Delivering the first workshop

The first workshop was the foundation on which the CS engagement was based, and involved a lot of information being passed onto the participants. The LLs had to introduce the iSCAPE project, their LL and its work, the use of CS in relation to air quality, and the SCK. It was also imperative that the workshop prepared the citizen scientists with a solid plan and structure for their 2 weeks of data collection on which the second workshop would be based.

The first workshop had a duration of approximately 3 h and had the following structure:

- *Warm-up exercise—Draw your air quality superhero* (5–10 min): The warm-up exercise consisted of drawing the participant's air quality superhero. This exercise helped to put the participants into an appropriate, open mindset and made them think differently about "how to control or tackle air-polluted areas within their city." As well as improving creativity, drawing helps facilitate improved memory, communication skills, and, most importantly, problem-solving skills.
- *Presentation introduction to CS* (15 min): This presentation was delivered by the local iSCAPE LL leader and was aimed at giving the audience a good understanding of what citizen science is and how this approach can deliver impactful contributions to the research work on improving air quality in cities.
- *Empathy-mapping effects of air pollution* (30 min): This team-building exercise is focused on gaining a better understanding of cities and citizens affected by air pollution. For CS workshops to work best, it was important that participants were engaged in the subject area, with some personal experience of air quality issues or awareness of specific examples/relevant research studies. Empathy mapping helped participants understand different situations and helped them empathize with those most affected by city air pollution, such as young children, older people, and people with asthma and other lung-related illnesses that impact breathing. The participants were also presented two specific research questions, the same questions in all six LLs.
- *Get to know your citizen kit* (20 min): This consisted of a hands-on session on how to use the iSCAPE SCK. The participants were divided into groups (the CS working groups), and a

sensor was given to each group. Then, the LL leader guided the participants on setting up the sensor and gave practical information on how to use it and collect data. An online documentation about the sensors was also presented.

- *Sensing checklist—Plan your data collection!* (30 min): The empathy-mapping exercise helped participants to gain a better understanding of specific situations, areas, and times when air quality impacts citizens. The chosen research question, and an understanding of the subject area, determined what kind of data collection (beside the sensors) the participants chose and what tools or methods to apply. The participants filled out the "Sensing Checklist", a form aimed at helping the participants to structure and plan their sensing activities. As such, it included reminders to take notes and photographs during their monitoring campaigns. It also asked to list the main barriers when collecting data and to sketch a draft timeline and map for the data collection.
- *Data advice—Plan your data collection!* (30 min): This section was designed as a follow-up to the previous activity. The teams presented their "Sensing Checklist" to each other. The facilitators shared tips about collecting information that can supplement the sensor data, e.g., taking notes, collecting feedback, recording sounds, or taking photographs. Best practices for successful data collection were also discussed, and practical tips for successful data collection were presented.
- *Closing of the session* (10 min): This session included final recommendations.
- *Questions and what is next* (15 min): This session was used to answer questions and also to give a quick outline of what would happen next. In specific, the LL lead explained how the team would provide support during the data collection phase and how the collected data would be analyzed and visualized, and provided some details on the content of the second workshop.
- *Gather feedback—Online and offline* (20 min): This session was used to collect feedback on the workshop in order to improve future workshops.

2.3 Data collection—Providing support during the data collection phase

After the first workshop, the CS teams or individual citizen scientists started collecting data. As well as collecting air pollution data, the citizen scientists were encouraged to take pictures, write notes, and talk to people in the area. Each LL started on a different day, which allowed for increased learning, iteration, and problem-solving between the LLs during weekly group catch-up calls. Most LLs experienced minor issues with the sensors, which were solved through individual catch-up calls with the partner leading the development of the sensors (Fab Lab Barcelona). Many participants also had minor difficulties during data collection, providing each LL a further touchpoint with their CS participants, during which the LL could provide support in uploading data, checking the suitability of the data, and answering upcoming/open questions.

2.4 Data analysis and visualization

Prior to the second round of CS workshops, a strategy for how to best analyze the data and create suitable visualizations that can be easily understood and used for the storytelling exercise in the second workshop was prepared. The iSCAPE data scientists used Python to analyze the air quality data and created two types of visuals using two types of data sets: noise (dBA) and PM 2.5 for every LL city. The two types of visualizations consisted of a time series and a heatmap presented in Fig. 1.

All data visualizations were created in Python, the codebase for which can be found on GitHub.[c] Running the analysis via this open source/GitHub platform is integral to the iterative evolution and development of sensor analysis frameworks[d]. Participants

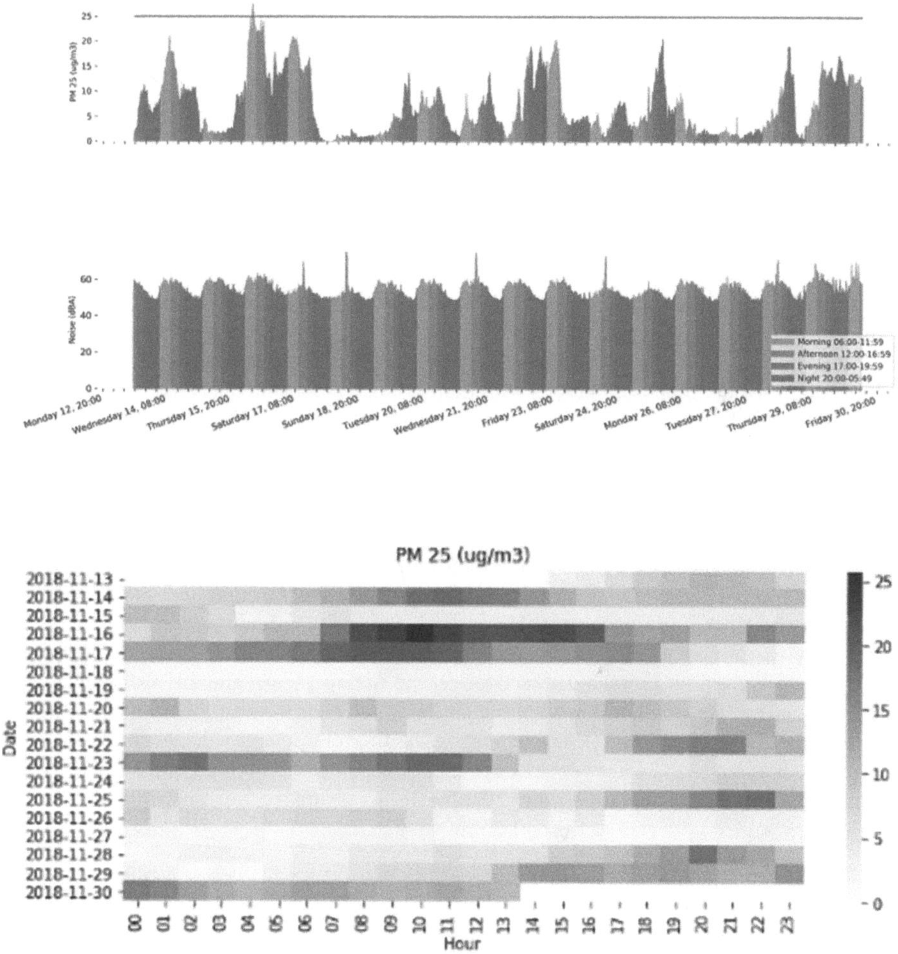

Fig. 1 Time series for PM 2.5 and noise (A) and heatmap for PM 2.5 (B).

[c] https://github.com/fablabbcn/smartcitizen-iscape-data/blob/Data_Visualisation_FCC/notebooks/FCC%20Timeseries%20Visualisation.ipynb.

[d] https://github.com/fablabbcn/smartcitizen-iscape-data.

are encouraged to contribute to fablabbcn/smartcitizen-iscape-data development by creating an account on GitHub.

An interesting finding from the data collection phase was that a direct correlation was observed between data accuracy and how well sensors were set up and managed.

2.5 Delivering the second workshop

For the second workshop, LLs welcomed citizen scientists back to view data visualization worksheets (which had been individually developed for each team of citizen scientists). This second LL workshop opened with a summary of how the data scientists had processed the citizen scientist data and why the specific visualizations had been created. The aim of the workshop was to demonstrate the meaning, value, and potential of the data the citizen scientists had collected.

The workshop lasted 2–3 h and followed this structure:

- *Introduction* (20 min): At the beginning of the workshop, short updates from data collection were given to the participants.
- *From raw to tangible data—Data awareness and analyzing data sets* (30–45 min): The data collected by the participants were analyzed and visualizations created by the iSCAPE team, as discussed in the following sections. The results were presented to the participants using the data visualization worksheet presented in Fig. 2. The CS teams then analyzed their data and compared the times and graphs with their notes and sensing checklist.
- *Storytelling through data—Developing a story and presenting* (30–45 min): This part of the workshop was devoted to engaging the participants in finding solutions to the measured air pollution. After the citizen scientists analyzed their data visualization, they started with their final exercise "Storytelling through Data—How to Make Your Data Tangible" (Fig. 3), where they had to create a short story of their research activities by using their data visualizations (as well as images and notes they took during the data collection phase). The worksheet asked the following question: "If you would have the opportunity to present your results to the city authority, how would you communicate your findings? Develop a convincing story that initiates change. Make use of all the data you collected." At the end of the workshop, the teams presented their stories and action plans. The LL facilitators were encouraged to take notes and prompt the audience for questions after each presentation.
- *Closing of the session* (10 min): The workshop was then closed with a discussion on the overall workshop work and some remarks about potential future activities.
- *Gather feedback—Online and offline* (20 min): Feedback on the workshop and on the sensing kits was then collected at the end of the workshop.

The second workshop was also used as an opportunity to discuss and plan next steps for the CS community and LL and in specific to start the planning of future engagement and facilitation of potential CS activities.

3 Results of the iSCAPE CS framework

All of the CS workshops in the six LL cities were successful and ran smoothly due to the comprehensive plan and framework structure. This section aims to summarize the outcomes from each workshop stage. The majority of LLs reported that the citizen

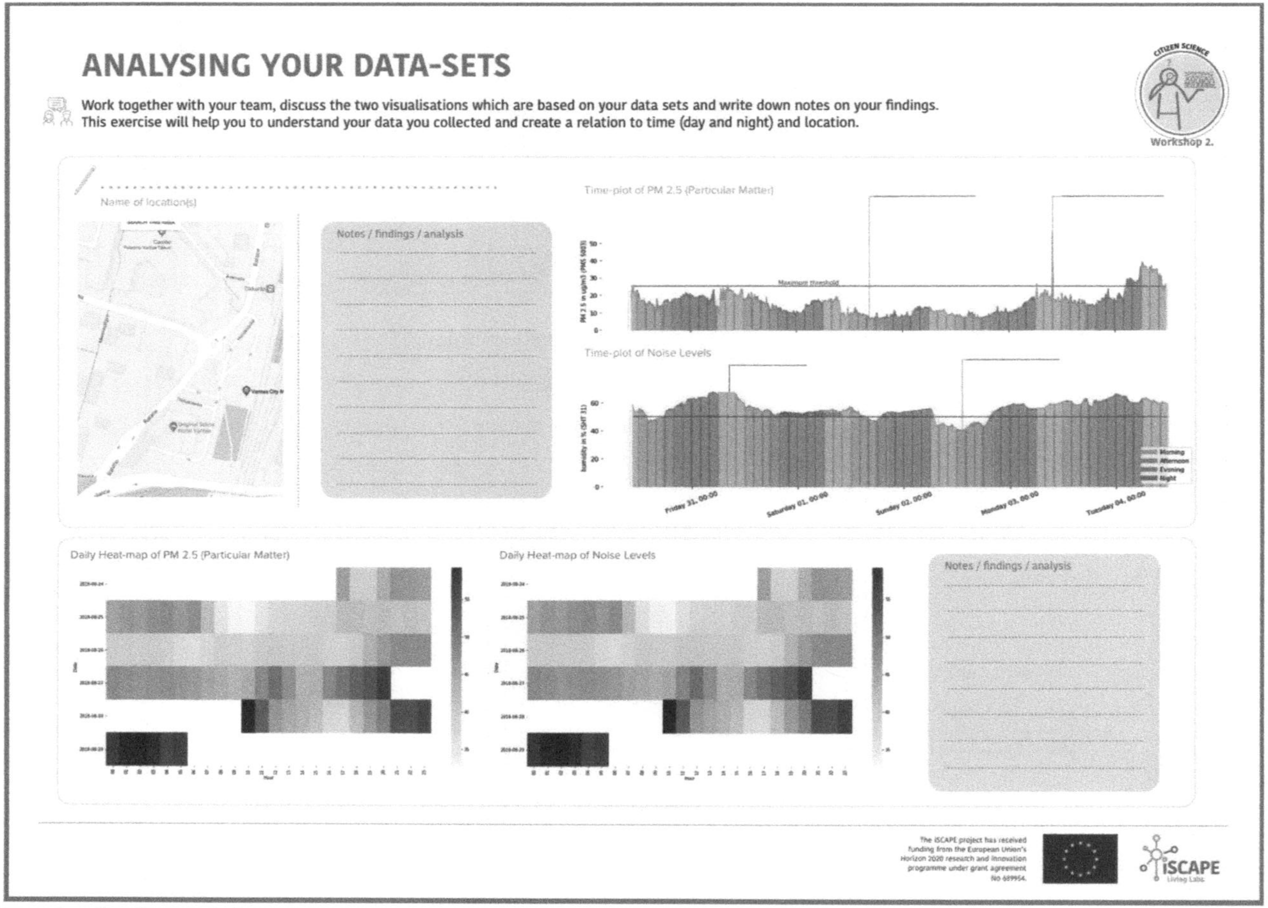

Fig. 2 Data visualization worksheet.

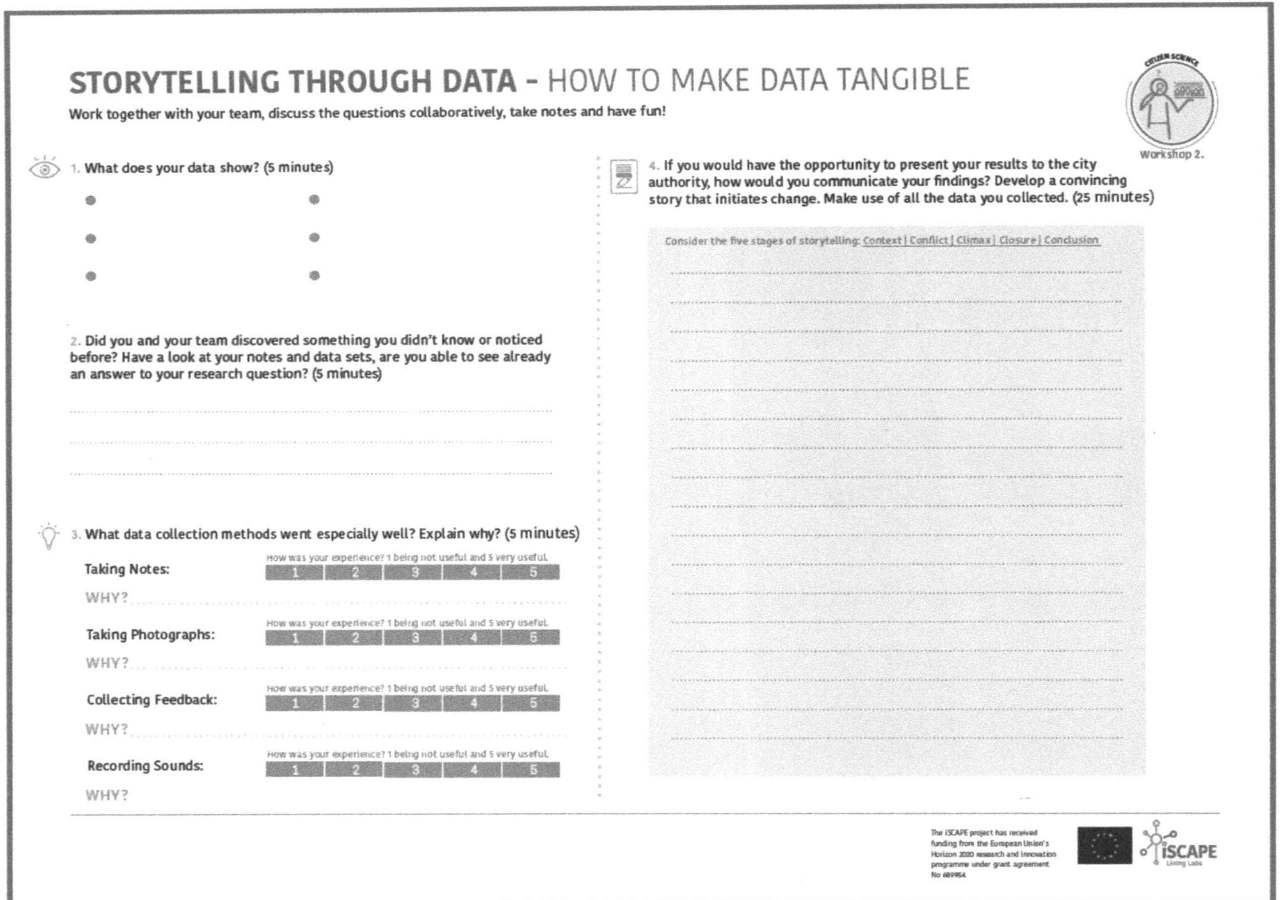

STORYTELLING THROUGH DATA - HOW TO MAKE DATA TANGIBLE

Work together with your team, discuss the questions collaboratively, take notes and have fun!

Workshop 2.

1. What does your data show? (5 minutes)

-
-
-
-
-
-

2. Did you and your team discovered something you didn't know or noticed before? Have a look at your notes and data sets, are you able to see already an answer to your research question? (5 minutes)

3. What data collection methods went especially well? Explain why? (5 minutes)

Taking Notes: How was your experience? 1 being not useful and 5 very useful. 1 2 3 4 5

WHY?

Taking Photographs: How was your experience? 1 being not useful and 5 very useful. 1 2 3 4 5

WHY?

Collecting Feedback: How was your experience? 1 being not useful and 5 very useful. 1 2 3 4 5

WHY?

Recording Sounds: How was your experience? 1 being not useful and 5 very useful. 1 2 3 4 5

WHY?

4. If you would have the opportunity to present your results to the city authority, how would you communicate your findings? Develop a convincing story that initiates change. Make use of all the data you collected. (25 minutes)

Consider the five stages of storytelling: Context | Conflict | Climax | Closure | Conclusion

The iSCAPE project has received funding from the European Union's Horizon 2020 research and innovation programme under grant agreement No 689954.

iSCAPE Living Labs

Fig. 3 "Storytelling through Data—How to Make Your Data Tangible" worksheet.

scientists were very engaged in the workshops and excited about the air quality monitoring activities (they often had prior knowledge of monitoring and air quality issues). Some citizen scientists came with clear goals, knowing what they wanted to get out of the iSCAPE CS activities. The LLs supported the citizen scientists as much as possible during the data collection phase (including uploading the data sets on the smartcitizen.me platform[e]) and resolved problems.

The Air Quality Superheroes warm-up exercise seemed a natural fit for people with a creative background and/or younger generations. The outcomes are fun, and the tool proved successful in encouraging conversations between citizen scientists. The results of this exercise do not directly influence the main CS workshop as its focus was to help everyone relax and feel at ease.

Empathy Mapping was the first team exercise. Citizen scientists often focused on vulnerable people such as children, the elderly, and those with asthma. Commonly imagined situations included a cyclist cycling to work close to the road, a bus stop next to a very noisy and polluted busy road, parents with a toddler in a pushchair, and pedestrians walking along a road. Exploring what could change to improve air quality, citizen scientist teams came up with options to avoid or minimize the production of air pollutants such as smart traffic lights, better traffic schemes, more public transport options, and more affordable electric vehicles. This exercise leads to the planning of their data collection (Sensing Checklist). The majority of citizen scientists developed very tangible hand-drawn maps, which were helpful when discussing the process of monitoring air quality and future collaborations. The Sensing Checklist helped the teams to frame their interest and allowed them to get an overview of the potential pollutants they could monitor.

Many citizen scientists came up with their own ideas and locations they wanted to investigate further. Upon analyzing the responses/stories, four research areas were identified that the citizen scientists were most enthusiastic about. As well as the two questions supplied, participants explored "indoor air quality vs. outdoor air quality" or just "indoor air quality", and "the balcony—examining the effect of smoking (smoking by neighbors on the balcony)." They also compared the effect of roads and wood burning on air quality. One team highlighted their story that their "data is based on real measurements monitored by a noncommercial authority." This was mentioned due to the habitual mistrust of figures due to data often being exaggerated in the press "for a good story."

3.1 Question 1: Hot spots

This research question focused on identifying air pollution hot spots in the six LL cities, where citizens would be most exposed to pollution. The citizen scientists observed pedestrian and bicycle ways and compared city locations with rural areas as well as compared peak hours on busy and less busy roads. They also monitored parking areas and roads close to schools. Their findings showed that the air quality fluctuates during

[e] https://smartcitizen.me/.

rush hour, but that pollution is often too high (PM 2.5), causing breathing problems for road workers. The air quality in rural areas overall is also strongly affected by roads and/or a nearby city center.

3.2 Question 2: Bus stop

This research question focused on assessing whether the location of a bus stop should take into account nearby traffic light systems because of increased exposure to air pollution. Citizen scientists compared different bus stops, monitored larger road crossings with traffic light systems, and bus stops close to parking spaces. Only a small number examined local bus stops in relation to traffic lights. However, one of the teams emphasized the importance of a sheltered bus stop, arguing that as people using public transport are already championing environmentally/sustainable modes of transport, they should not be exposed to more pollution as a result. Measurements showed that, especially during peak hours, bus stops have high noise levels and alarming levels of PM 2.5.

3.3 Question 3: Indoor air quality vs. outdoor air quality

CS teams initiated their own research focusing on indoor air quality. Most participants were very surprised about the indoor air pollution levels, which were often worse than the air outside. They monitored different rooms—kitchen (whilst cooking), bedrooms, and the living areas—and compared indoor and outdoor measurements. They also monitored their work location (inside an office) focusing on both air and noise pollution. The team that analyzed workplace air quality identified PM levels more than four times the average. Possible sources of PM were local road/trains, industrial location, and/or the local printer.

3.4 Question 4: Balcony location

This research question focused on exploring the effect of smoking and close-by traffic using balconies as monitoring locations. CS teams investigated factors that contribute to pollution on their balcony as well as living room. They monitored local traffic time from peak hours to the weekend and measured air quality when neighbors were smoking on the next-door balcony, as well as comparing times and PM levels when people start heating their homes or, in Finland, use the sauna.

4 CS workshop findings and reflections

The findings from the CS workshops were extracted from the reflection sheets and the participant feedback.

Whilst the two research questions created a focus for the citizens, they also allowed the LLs and a managing partner to observe patterns across the six European cities.

Participants were eager to develop their own research questions and often discussed research questions or locations where they could measure air pollution. Across the six cities, participants asked many technical questions, including health-related questions. Only some citizen scientists had prior technical knowledge or experience (e.g., those at Fab Lab, Bottrop, Germany, and Vaisala Inc. and the Environmental Service (air quality focus), Vantaa, Finland). Despite a lack of technical expertise, participants across the six cities were eager to learn more and use the citizen kits.

All participants and facilitators gave positive feedback about the data visualization worksheet: it generated active and meaningful discussions and the citizen scientists were excited to see their research/data being visualized. A keen interest was observed in all cities with respect to understanding the peaks and anomalies in the graphs/ visuals.

> *Indoor measurements created a great disappointment in the teams, due to the surprisingly bad air quality results (LL, Bologna).*

LLs shared fantastic stories developed in the second workshop. However, some LLs mentioned that while some participants struggled with storytelling (unable to find ideas for the story or the right words), others felt that they did not have sufficient data to tell a story. As with the first workshop, participants asked a lot of technical questions. Encouragingly, in some LLs, participants were very interested in the feedback from other citizen scientists.

> *The difference in the second workshop was that the workshop results changed the views about vehicles, domestic indoor pollution, and emissions (LL, Guildford).*

Some useful feedback was also collected on how to improve the overall iSCAPE CS framework. It appeared important to have a technical expert present in person or remotely by phone or videoconference to address in real- time the initial issues with setting up the sensors. Another important point that was raised was the need to set up a meeting point for the citizen scientist and the facilitators (LL) to get support for data uploading and fixing of small technical issues. This should also be complemented with a more detailed sensor explanation, including a run-through of quick fixes.

The participants were also extremely keen to elaborate their own research questions, pointing to the potential of this kind of CS workshops to give raise to a grassroots initiative around the theme of air pollution and to empower citizens to take action to contribute to more sustainable cities. As such, workshops should also focus on self-initiated research questions to foster and facilitate bottom-up initiatives. The research questions allowed for similarities in the data sets to be highlighted, and sped up the decision-making process during the workshop; the space made for broader self-initiated inquiry should, however, be carefully considered to ensure the validity of any further comparisons.

Another important element that emerged from the six LLs was the importance of understanding the audience. LLs need to get to know their audience, which will help them to cater to and utilize the citizen's needs, expertise, and interests. Recently, the LLs made the first steps toward creating a CS community, so that future CS activities and interaction with the public will refine the skill. Such expertise will influence the interaction between different groups of citizen scientists; for example, some like to be guided, whilst others like to create their own projects and prefer a check-in from time to time. The LLs need to find their way of collaborating or facilitating with citizens on air quality monitoring activities. Being aware of the diversity of people (e.g., different age groups, disabilities, or citizens with different backgrounds) and how their needs differ, can only be found out when delivering a larger number of CS workshops with diverse audiences.

Throughout the LL activities, keeping momentum and motivating the public to engage in CS activities is a lot of hard work and a long process. The "right" methods for engagement need to be developed over time, as more is known about the audience.

In general, a common outcome of the CS workshops across the six LLs was that they gave rise to bottom-up monitoring initiatives and a general sense of empowerment for the involved participants. Some LLs reported that participants are planning to use iSCAPE CS results for their own initiatives, e.g., in Vantaa, a citizen scientist wants to use their results to make a case to stop neighbors (in their apartment building) from smoking on balconies. In the other LLs, some motivated citizen scientists requested for the use of low-cost sensors, outside of the CS workshops, for specific purposes or research projects so that they could monitor longer and/or different locations. This is a very interesting finding as it shows that citizens, if empowered with appropriate knowledge and tools, are willing to carry out and also develop activities related to improving air quality in their cities.

Moreover, the CS workshop enabled the iSCAPE team to road-test the Citizen Kit under real, practical conditions. This, of course, meant actual deployment and testing of all the individual components of the proposed end-to-end solution—from the sensor hardware, the onboarding website, and the data visualization platform. Whilst some of the elements—for example, the data platform—have been tested before, this more wide-ranging and testing deployment meant that they could be improved iteratively over the course of the iSCAPE project. It was particularly useful in validating and improving the software that ran on the sensor devices. The opportunity to run the software under different practical deployment conditions—e.g., users with more or less technical skills and variable wi-fi connections—helped improve corrections and precision. All of this helped inform the creation of the commercial version of the SCK, the Citizen Kit 2.1.

5 Conclusions

Each LL highlighted the advantages and positive effects CS can have on a city, research project, policy design, and the citizen scientists themselves. Combining a campaign approach with community-centric access to research technology and guidance (low-cost sensors and CS activities) allowed citizens to get involved and encouraged the first steps toward change. The green shoots of the longer-term goal of establishing a self-sustaining CS community were observed in a number of instances. The behavioral changes also encouraged citizens to reduce their own contribution to pollution.

As well as these behavioral changes that encouraged citizens to demand informed changes from their local and national governments in terms of environmental policy and action, the iSCAPE CS work is aimed at facilitating further collaboration with the local authorities, municipalities, community groups, relevant organizations, and other universities/academic institutions through the six LLs. Changing policies is a long-term process that needs strong collaborative relationships with the city and solid evidence. In this context, it was evident that the benefits, both perceived by the participants and actual, of the CS activities were part of a wider project and ecosystem. In specific, the fact that the CS workshops were part of the LL activities and of a multicountry European project made the participants feel to be part of a wider scheme and produced also the actual benefit to engage a wide range of stakeholders in the LL activities. Each LL has strong links with its municipality and with a wide range of local and national stakeholders, allowing for a higher potential impact of these bottom-up citizen science activities.

The project is a great demonstration of the "think global, act local" maxim put into practice. Whilst the issues and concerns are affecting the whole world, the motivation for change is driven through citizen engagement and empowerment of communities at a local level. As an example, the CS activities in Dublin have attracted the interest of other communities and national bodies, including the Irish Environmental Protection Agency. In Vantaa, the "Living Lab" and "Citizen Science" ethos are now being used in other projects such as the ongoing H2020 OPERANDUM project, where the Open Air Laboratory uses LL interactions with citizens and stakeholders. Projects such as iSCAPE and Making Sense[f] clearly demonstrate how participatory citizen engagement projects foster awareness. By giving participants the capacity to "sense" the pollutants impacting their environment, they are empowered to get involved in a practical, hands-on approach. In this important way, they are shown to adopt a concern about and ownership of the issues.

The experience gathered during the CS workshops and activities in the six iSCAPE LLs exposed several advantages and also room for improvement for the framework used. It was clear that these activities with such a variety of users were extremely beneficial to fine-tune the low-cost sensing kits and solve several bugs and issues. The impact on the awareness of the citizen scientists about air pollution was extremely positive and was increased by the hands-on experience with low-cost sensors. Even though citizen scientists had some difficulties with low-cost sensors, they learnt to make the best out of the situation and worked with what they had. Over time, they also learnt how to use the sensors more effectively and came up with solutions for common bugs. The results show the clear value of involving and supporting motivated "citizen scientists." The process of experimentation and testing has also contributed directly to the next iteration of the SCK and scheduling of a number of follow-up projects. The main benefit of the interaction between citizens and low-cost sensing units

[f] http://making-sense.eu/.

was in terms of empowering the participants with the capabilities and tools to contribute to the research in the area of air pollution control. It allowed an increase in the motivation of the participants because they all felt they could actually make a change and contribute to more sustainable policies by having access to instruments to make more informed choices.

Reference

Camprodon, G., González, Ó., Barberán, V., Pérez, M., Smári, V., de Heras, M.Á., Bizzotto, A., 2019. Smart Citizen Kit and Station: An Open Environmental Monitoring System for Citizen Participation and Scientific Experimentation. HardwareX, p. e00070.

The role of environmental specimen banks in monitoring environmental contamination

6

Jacqueline S. Chaplow[a], *Alexander L. Bond*[b], *Jan Koschorreck*[c], *Heinz Rüdel*[d], *and Richard F. Shore*[a,†]
[a]UK Centre for Ecology & Hydrology, Lancaster, United Kingdom, [b]Bird Group, Department of Life Sciences, The Natural History Museum, Tring, United Kingdom, [c]Federal Environment Agency, Berlin, Germany, [d]Fraunhofer Institute for Molecular Biology and Applied Ecology IME, Schmallenberg, Germany

1 Introduction

The term *environmental specimen bank* (ESB) refers to an organization or facility that collects, preserves, and stores environmental specimens. These specimens may have been collected as part of a systematic monitoring program, produced from a scientific research study, or form part of legacy collections that have subsequently been donated to the ESB. Targeted collections may have been conducted to sample a specific habitat or address an environmental issue. Overall, specimen collections can result from short-term discrete sampling projects, long-term research or monitoring programs, or systematic collections or can be a historical archive.

Specimen banks, and the environmental samples they contain, are a key tool that enables examination of how the environment is affected by, and responding to, major long-term anthropogenic-mediated changes. Human population growth has driven the demand for land, food, water, energy, and manufactured goods. This demand has led to land-use change, habitat and biodiversity loss, global climate change, and wide-scale industrial pollution (Luo, 2018). ESBs have been linked particularly with the latter. For example, they have been used to track pollution and assess the effects of restrictions or bans on the use of potentially toxic compounds, such as legacy persistent organic pollutants (POPs). This has involved analyzing a time series of specimens to determine whether contaminant concentrations have fallen following the implementation of such mitigation (examples are given in Section 5). ESBs can also be used to identify the emergence of new environmental contaminants. According to the European Inventory of Existing Commercial Chemical Substances, more than 100,000 different compounds are produced each year, and between 1000 and 2000 new products enter into circulation annually (Vázquez et al., 2007). Meanwhile, Koizumi et al. (2009) estimated that the environmental fate of only a few of the estimated 70,000

† Deceased.

Monitoring Environmental Contaminants. https://doi.org/10.1016/B978-0-444-64335-3.00002-5

chemicals that are commonly used in industry have been characterized. Identification of the actual environmental presence and the fate of these compounds and/or their breakdown products, and how these vary spatially and over time, can often require analysis of samples that have been stored in the ESBs.

The International Environmental Specimen Bank Group (IESB) promotes the worldwide development of techniques and strategies for environmental specimen banking. Information on the IESB website (http://www.inter-esb.org/) includes a list of countries maintaining environmental specimen bank activities, archiving protocols from the German, Swedish, and US Marine ESBs and a list of ESB workshops, meetings, and conferences (some international) held, and their proceedings since 1991. More than 20 international ESB meetings and special ESB sessions at conferences have been held in different countries since 1977 (Zhao et al., 2015).

Across the world, there are a great number of different specimen banks that function in a wide variety of ways. Their remit may be broad or narrow, with sampling ranging from worldwide to small, local catchments. Some ESBs take a wide variety of environmental specimens, while others have a specific focus or take a narrow spectrum of samples from a designated sampling area on a strictly temporal basis. In this chapter, we aim to bring the wealth of information about ESBs up-to-date.

2 A general overview of environmental specimen banks

ESBs have been established for various reasons. Many were set up to provide material for the examination of long-term trends that required retrospective analysis of environmental samples (Poulsen and Pritzl, 1993) using new methods. Others were developed to function as systematic or synoptic collections. Across the world, there are a great number of specimen banks that function to fulfill the following roles:

- To provide samples that can be used to evaluate wildlife and domestic animal health and human food safety
- To investigate trends in environmental change over time
- To detect and identify sources of legacy and newly emerging chemicals
- To validate computer model predictions and test novel technologies
- To test and evaluate government environmental policy and regulations
- To serve as archives for environmental specimens and facilitate specimen use

The first specimen bank was established in Sweden in the 1960s (Tanabe and Subramanian, 2011; Koizumi et al., 2009). Its aim was to study pollutant residues and their effects on the environment. The number of ESBs has since grown, and there are currently more than 30 across the world. A selection is listed in Table 1, but this does not include natural history collections in museums, of which there are hundreds globally (Green and Scharlemann, 2003; Roselaar, 2003). Many museums have restrictions on what can be sampled and how, because their primary function is seldom to provide an archive for chemical analyses, but such sources of material can, in some cases, be the only opportunity to investigate changes over longer time periods (e.g., Vo et al., 2011; Bond et al., 2015).

The scope and breadth of the existing national ESBs vary between countries. Many specimen banks archive specific samples (e.g., human tissue or marine mammals) or

A selection of environmental specimen banks from across the world.

Country	Name of ESB	Approximate start	A broad description of the types of specimens archived
Canada	National Aquatic Biological Specimen Bank	1977	Fish and plankton
	Canadian Wildlife Service Specimen Bank	1963	Bird, mammal, fish, reptile, amphibian, and invertebrate
China	Shanghai Academy of Environmental Sciences and Shanghai Institute of Nuclear Research	1998	Water, sediment, and dust
	Polar Research Institute of China	1989	Polar flora and fauna, ice, rock, and sediment
	China Ocean Specimen Bank	2001	Cobalt-rich crusts, polymetallic nodules, and hydrothermal sulfides from ocean surveys
	Yangtze Environmental Specimen Bank (YESB)	2011	Plant, animal, and human (from the Yangtze River Basin)
Denmark	Faroe Island Environmental Specimen Bank	1993	Plant, invertebrate, and vertebrate
	Tissue and Data Bank for Greenland	1968	Pathological
Finland	Paljakka Environmental Specimen Bank	1994	Moss, lichen, bark, humus, and forest litter
France	Observatoire de Recherche sur l'Environnement (ORQUE)	2001	Environmental
	ANDRA Observatoire Perenne de l'Environnement (OPE)	2010	Radioactive waste
	Mytilothèque	1979	Shellfish
Germany	Umweltprobenbank at Federal Environment Agency	1982	Animal and plant samples from terrestrial, limnetic, and coastal ecosystems (e.g., mussels, fish, bird eggs, and tree leaves), and human samples
	Environmental Specimen Bank for Human Organ Specimens		Human
Italy	Italy Mediterranean Marine Mammal Tissue Bank	2002	Marine mammal
	The Antarctic Environmental Specimen Bank (BCAA)	1994	Marine biota and sediment

Continued

Table 1 Continued

Country	Name of ESB	Approximate start	A broad description of the types of specimens archived
Japan	Japanese Environmental Specimen Bank for Global Monitoring (es-BANK)	1966	Bird, fish, reptile, amphibian, mammal, and invertebrate
	Time Capsule Program for Environmental Specimens (NIES ESB)	1979	Fish, shellfish, marine sediment, airborne particles, and human breast milk
	Kyoto University Human Specimen Bank	2004	Human
Norway	Norwegian Environmental Specimen Bank	1993	Fish, shellfish, terrestrial and marine bird and mammal, plant, air, and sediment
South Africa	Biological Resource Bank (BRB)	2005	Wild animals
Spain	The Biscay Bay Environmental Specimen Bank (BBEBB)	1980s	Fish, marine mammal, mollusk, and shark
	Environmental Specimen Bank of Galicia (ESBG)	2001	Macroalgae
South Korea	National Environmental Specimen Bank (NESB)	2007	Fish, bird eggs, and trees
Sweden	Swedish Environmental Specimen Bank	1960s	Marine mammals and birds of prey
United Kingdom	National Fish Tissue Archive	2000	Fish
	Predatory Bird Monitoring Scheme Tissue Archive	1960s	UK raptors and owls
	Cardiff University Otter Project	1992	Eurasian otter
United States of America	Marine Environmental Specimen Bank	2001	Marine mammals, shellfish, fish, bird eggs, and feathers
	National Biomonitoring Specimen Bank (NIST)	1979	Human, marine sediments, fish, shellfish, and marine mammals
	Alaska Frozen Tissue Collection	1991	Marine mammals

samples from specific locations (e.g., a catchment or forest), while others archive a wide range of samples from multiple habitats and ecosystems. In addition, sampling may be according to a strict regime (e.g., annual or 5 yearly), in response to a specific event (e.g., chemical release), or may develop randomly. We have identified a small selection of specimen banks to outline this range:

- The Canadian Wildlife Service Specimen Bank archives samples from a range of Canadian habitats, whereas the Paljakka Environmental Specimen Bank (Finland) was set up far from anthropogenic emission sources to investigate environmental change in the Paljakka forest.
- The Japanese es-BANK (http://esbank-ehime.com) archives samples from all over the world, whereas the environmental specimen bank of Galicia only stores aquatic bryophytes collected in the Galician river network of northwest Spain (Vázquez et al., 2007).
- The Umweltprobenbank (https://www.umweltprobenbank.de (German ESB)) collects and stores limnetic, marine, terrestrial, and human samples collected from designated sampling areas in Germany every 5 years.

The Swedish, German, American (at the National Institute of Standards and Technology (NIST)), and Norwegian ESBs are regarded to be pioneers in the field of specimen banking (Marigomez and Lekube, 2017). Due to continuous funding and a systematic sampling approach, the German ESB is generally considered to be the gold standard of specimen banking.

3 The German ESB—The Umweltprobenbank

The German ESB provides the scientific basis for decisions made by its funder, the German Federal Ministry for the Environment, Nature Conservation and Nuclear Safety (BMU), on environmental protection measures. The German ESB is a permanent monitoring facility that investigates the sources, environmental fate, and changes over time of environmental contaminants, using both environmental and human samples. Its main tasks are long-term storage of environmental and human specimens and provision of a continuous historical record of the state of the environment in Germany. The German ESB also responds to requests for access to samples.

Operations are managed by the German Environment Agency (Umweltbundesamt), and specialist research institutes are contracted for specific tasks (sampling, sample preparation, storage, and analysis). Sample collection of human blood began in 1982, followed in 1985 by the collection of limnetic samples (suspended particulate matter, zebra mussel and bream in rivers and lakes), marine samples (common bladder wrack (brown algae), blue mussel, eelpout, and herring gull), terrestrial samples (common spruce, pine, poplar, beech, roe deer, pigeon, earthworm, and soil), and further human samples (blood and urine (and, until 2004, saliva and hair) from students aged between 20 and 29 years).

Routine operation of the German ESB started in 1985 (Gies et al., 2007; www.umweltprobenbank.de/en/). A sampling program was initiated to allow for the development of a comprehensive database to assess the state of the national environment over time and to support chemical risk assessment by providing exposure data, especially for biota.

The ESB operates by following a concept document that describes the principles of sampling, sample processing, and long-term storage (UBA, 2008). This covers aspects such as the selection of representative sampling areas and specimen types as well as sampling procedures, sample characterization, transport, storage, and documentation. Every step (sampling, transport, preparation, analysis, and long- term storage) is subject to binding standard operating procedures (SOPs) for each specimen type. An example of the type of information in a typical ESB protocol is given in Table 2. The use of stringent SOPs ensures good quality assurance for the sampling and storage processes and hence consistent operation over past and future decades. This consistency lends confidence to the scientific information and conclusions drawn from studies on the samples.

Sample storage involves freezing samples as soon as possible after sampling. They are stored in the gas phase above liquid nitrogen and are kept below the glass transition temperature of water (about −135°C). This prevents all physical processes (e.g., ice-crystal growth) and minimizes chemical processes. The inert atmosphere from the evaporating liquid nitrogen prevents contact with oxygen and, therefore, oxidation processes.

Samples (1–2 kg of sample mass) of each of the specimen types that form part of the core collection of the Umweltprobenbank are collected annually. They are homogenized by cryomilling at temperatures below −135°C under clean-air conditions (achieved using particle and activated charcoal filters). This low temperature makes biota samples brittle so that they can be easily crushed and ground (Rüdel et al., 2008). For example, the contents of 70 herring gull eggs from a single site are used for an

Table 2 Example of the details contained in an Umweltprobenbank SOP.

Protocol	Fish sampling
Species	Common bream (*Abramis brama*)
Guideline	Klein et al. (2018)
Rationale for selection	Widespread presence, abundance at sampling sites and substantial biomass availability
Tissues stored	Muscle and liver
Sampling period	Late summer after spawning
Species determination in the field	Description of typical features that allow unambiguous identification
Biometric measurements to be taken	Body length, body weight, and sex
Number of samples to be taken	20
Dissection	Filet and liver dissected on a clean bench in a mobile laboratory

annual pooled sample, and, likewise, approximately 20 freshwater fish filets are bulked to form a pooled sample. After cryomilling, the particle size of the milled sample is measured by laser diffraction, and more than 90% of most materials have particle sizes of <200 μm. This low size and thorough homogenization ensure that aliquots of the pooled ESB sample are consistent in composition and representative of the whole bulk sample. Typically, up to 200 subsamples (5–10 g) are prepared from each annual sample and stored in the ESB cryoarchive, thereby facilitating a similar number of separate analyses. The resultant data contribute to the knowledge about the sample and are publicly available (https://www.umweltprobenbank.de/en/documents/investigations/analytes).

ESB protocols also cover the chemical analyses of ESB samples (for example, see Rüdel et al., 2011a). This covers all aspects of the analysis, ranging from sample extraction/digestion protocols, analytical techniques, and quality assurance (QA) protocols (preparation of standards, instrument calibration, avoidance of spectral interferences, and use of blanks and certified reference materials). Four subsamples are routinely analyzed. Mean values and standard deviations are reported, together with limits of quantification and QA information. Retrospective monitoring data from standardized ESB archive samples are already available for a large number of compounds. These include perfluorinated compounds in coastal samples (Rüdel et al., 2011b), persistent organic pollutants in gull eggs (Fliedner et al., 2012), EU Water Framework Directive priority substances in freshwater fish (Fliedner et al., 2016), novel brominated flame retardants in tree samples (Dreyer et al., 2018), and rodenticides in freshwater fish (Kotthoff et al., 2019).

Although several ESBs have based their operations on the Umweltprobenbank, many differ from the German model in a variety of ways, including the following:

- Facility type (purpose-built, deals only with ESB samples, or a freezer used by lots of different projects in a research department)
- Funding model (government department or agency, environment or wildlife charity, or private enterprises)
- Sample collection model (time-constrained research program, in response to citizen science, legacy samples, and museum collection)
- Storage (full samples retained long term, homogenized subsamples retained for a set period, and tissue preserved for the duration of study)
- Preservation state (frozen at a range of temperatures, under liquid nitrogen, desiccated, using a chemical (formalin, alcohol))
- Accessibility issues (permission, health and safety issues, and limited sample quantity)

4 Compliance with standards

It is essential that, where possible, ESBs use the most up-to-date and relevant standards for sample collection, processing, and analysis of specimens and samples and also for storing information. A selection of relevant standards is presented in Table 3.

Table 3 Examples of international standards applicable to the operation of ESBs.

General International Standard Archival Description	Adopted by the International Council on Archives Committee on Descriptive Standards (ICA/CDS https://www.ica.org/en) in 1999 (ISAD(G), 2000) • It is a set of seven descriptive elements designed to be used with existing national standards (or as the basis for development of standards) to identify and explain the context and content of archived material to promote accessibility
The Collections Trust (https://collectionstrust.org.uk/)	An independent UK-based charity helping cultural organizations to unlock the potential in their collections • Intelligence Hub provides the Spectrum Standard (https://collectionstrust.org.uk/spectrum/spectrum-5/), a collection management standard first published in 1994 • The standard is used by more than 7000 museums, galleries and cultural organizations worldwide
Darwin Core (DwC) (https://dwc.tdwg.org/)	A community standard for biodiversity data used to catalogue biological specimens, mainly by natural history museums • DwC includes a glossary of terms intended to allow the sharing of information about biological diversity (Wieczorek et al., 2012) • DwC primarily based on the occurrence of taxa as documented by observations, specimens, samples, and related information • DwC is coordinated by the Global Biodiversity Information Facility (GBIF) (https://www.gbif.org). This international network and research infrastructure provides data-holding institutions around the world with common standards and open-source tools that enable them to share information about where and when species have been recorded

Table 3 Continued

The International Society for Biological and Environmental Repositories (ISBER)	• The society publishes the journal *Biopreservation and Biobanking* • It champions the development of standardized methods for collection, long-term storage, retrieval, and distribution of specimens to enable future use of high-quality biological and environmental research specimens (Campbell, 2012; Campbell et al., 2018) • ISBER publications are periodically reviewed, revised, and published (i.e., 2005, 2008, 2012, and 2018) to reflect advances in research and technology

5 Examples of the use of ESB samples or data in scientific research

There are various examples that highlight the use of archived specimens in scientific research or for monitoring purposes. Here, we briefly present two case studies that demonstrate the use of both ESB samples and samples held in Natural History museums. The case studies highlight how interrogating specimens in ESBs and natural history collections can identify long-term, large-scale trends in environmental pollution and demonstrate the impact of environmental policies over long timescales. Such assessments would not have been possible without access to the archived samples.

5.1 The use of ESB archived raptor tissues to track changes in environmental pollution

Predatory bird tissues have been widely used to track changes in environmental contaminant concentrations at a variety of spatial scales (Gómez-Ramírez et al., 2014; Espín et al., 2016). Such work often requires long-term monitoring and an associated development of a tissue archive or a specialized ESB. The Predatory Bird Monitoring Scheme (PBMS, https://pbms.ceh.ac.uk/) is one such example (Walker et al., 2008). The PBMS is a long-term, national-scale UK monitoring scheme that quantifies the concentrations of contaminants in the livers and eggs of selected species of predatory birds in Britain. As part of its operations, it has developed a long-term tissue archive, the aims of which are to facilitate studies on chemical fate and behavior, support a wide range of conservation-related studies on raptors, and trial new monitoring. In 2014, the archive was estimated to contain close to 50,000 tissue and egg samples

(Walker et al., 2014), and approximately 2500 samples (tissues from some 300 carcasses and 100 eggs) are added to the collection annually.

Historically, the PBMS tracked declines in environmental contamination related to organochlorine pesticides, such as DDT and dieldrin. The tracking involved measuring the tissue concentrations of these compounds in sentinel species such as the peregrine falcon (*Falco peregrinus*) and the sparrowhawk (*Accipiter nisus*) (Ratcliffe, 1980; Newton and Wyllie, 1992). Declines in tissue dieldrin (HEOD) and DDE (a metabolite of DDT) residues were observed following the implementation of voluntary and then mandatory bans on the use of these pesticides (Fig. 1), and population numbers began to recover. The tissue residue measurements were supplemented by the measurement of eggshell thickness. This was because DDT caused thinning (and subsequently induces breakage) of eggshells. The monitoring demonstrated that eggshell thickness had decreased during the period when DDT was in use and then slowly increased following the ban on DDT, although recovery to pre-DDT levels took many years. Knowledge of the pre-DDT eggshell thickness values was obtained by examining eggshells that were held in museum specimens that dated back to the mid-19th century (Ratcliffe, 1980; Newton, 1986). Thus, ESB- and museum-archived specimens were both integral components of these studies on the impact of organochlorines.

Retrospective studies using the PBMS archive samples have also been used to demonstrate how environmental concentrations of polybrominated diphenyl ethers (PBDEs), used as flame retardants, have varied over time. The rise of contamination and the presence of supposedly nonbioaccumulative PBDE congeners in the biota were both demonstrated by examining the tissues and eggs in a range of species (Leslie et al., 2011). The results informed policymakers of the environmental risk posed from these compounds. Furthermore, concentrations in gannet (*Morus bassanus*) eggs clearly charted not only the rise in environmental PBDE concentrations but also their subsequent decline following increasing restriction on manufacture and

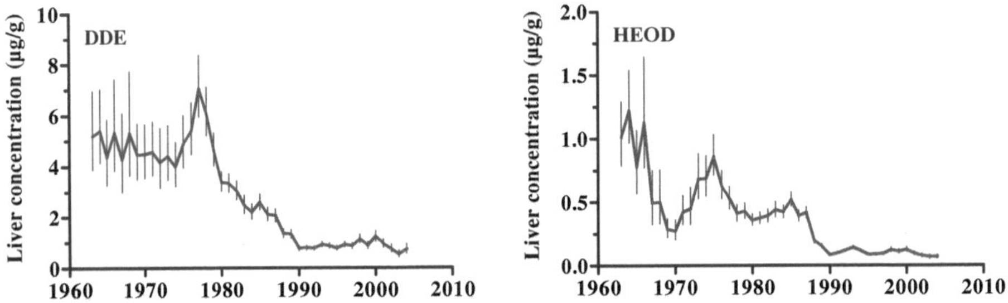

Fig. 1 Trends in pollutant residues in livers of sparrowhawks, 1963–2002. Data are three-year moving geometric means with one geometric standard error on either side and are wet weight concentrations.

Data are from Shore, R.F., Walker, L.A., Turk, A., Wienburg, C.L., Wright, J., Murk, A., et al., 2006. Wildlife and Pollution: 2003/04 Annual Report. JNCC Report, No. 391. Joint Nature Conservation Committee, Peterborough, UK, pp. 91.

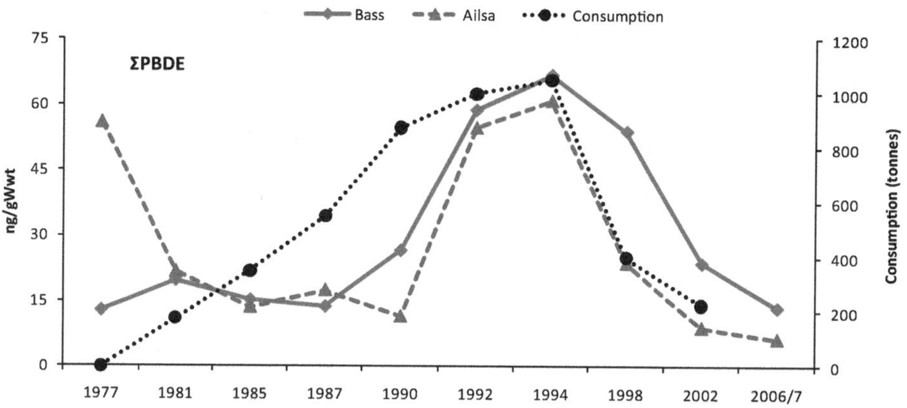

Fig. 2 Geometric mean concentrations (ng/g wet weight) of ΣPBDE in gannet eggs from Bass Rock (—◆—) and Ailsa Craig (--▲---) in comparison to median values of seven modeled estimates of consumption (••●••) (Prevedouros et al., 2004).
From Crosse, J.D., Shore, R.F., Jones, K.C., Pereira, M.G., 2012. Long term trends in PBDE concentrations in gannet (*Morus bassanus*) eggs from two UK colonies. Environ. Pollut. 161, 93–100.

use. The changes in egg concentrations closely tracked modeled production volumes (Fig. 2).

Both of these examples involved extensive use of ESB specimens. Many of the principles governing the use of samples from ESBs equally apply to the use of samples sourced from natural history collections. Museums may, in fact, store the same sample types as ESBs in some cases, particularly, sample types that do not require particular storage conditions—feathers are a prime example. Indeed, there is interest in determining the potential for how such samples, sourced from both museums and ESBs, might be used in combination to investigate long-term pollution trends (Movalli et al., 2019). Two examples of how archived feathers have been used to date to track pollution are given in the following case study.

5.2 The use of archived feathers to track changes in pollution over centennial timescales

While the earliest ESBs began operating in the 1960s, natural history collections contain specimens dating back to the late 18th century, making them the only opportunity to examine changes over more than 100 years. Two examples are analysis of feathers for methyl mercury, and whole specimens for black carbon.

Ivory gulls (*Pagophila eburnea*) are High Arctic seabirds that scavenge on marine mammal carcasses. Their population in Canada has declined precipitously since the 1980s, and contemporary studies have found that they have the highest concentrations of mercury in the eggs of any Arctic seabird (Gilchrist and Mallory, 2005; Braune et al., 2006). This raised the question of whether mercury exposure has always been high in these apex predators or whether high mercury concentrations are a recent phenomenon. Bond et al. (2015) addressed this question by examining mercury in the feathers of ivory gull specimens available in museum collections. Bird feathers are inert once grown, and after study skins are prepared, they can last hundreds of years. While there may be some post-collection chemical contamination of feather material, this can generally be accounted for through selective analyses of contaminants. In the case of mercury, this can involve analysis of methyl mercury, which would be present only because it would have been biologically incorporated while the animal was living, rather than total mercury, which may also be present as a preservative contaminant. The investigators found significant increases in feather methyl mercury concentrations from 1877 to 2007 without concomitant changes in the trophic level of the animals as determined by stable isotope analysis, which can also be conducted on feathers. The measured concentrations approached those thought to have deleterious effects and it was concluded that mercury may have been contributing to the species' decline.

While mercury is bound chemically to feathers, museum specimens can also be useful for understanding external contamination. DuBay and Fuldner (2017) measured reflectance of plumage color in five passerines from the US Manufacturing Belt between 1880 and 2015 and found that black carbon, an aerosol from incomplete petrochemical combustion and major component of soot, that had been deposited on birds while alive was related to coal combustion before the Clean Air Act (1963). After the introduction of regulations, black carbon on birds was found to have declined markedly, demonstrating the effectiveness of the regulatory intervention.

6 Future perspectives of ESB functions

Since the first ESB was set up in Sweden in the 1960s to complement the advent of monitoring studies, the need for, and functioning of, environmental specimen banks has developed. ESBs are now recognized as an essential component of environmental science research and are regularly used to provide samples (including legacy samples) to detect emerging chemical threats and chemical contaminants in the environment, to test novel methodologies or instrumentation, and to monitor human or environmental exposure trends. In future, there may be technological breakthroughs that allow for the analysis of newly relevant substances and the identification/analysis of complex mixtures of chemicals. ESB samples may be used for retrospective analysis and discovery of currently unknown compounds (Togola et al., 2017) and in nontarget screening, where all compounds can be detected and identified without any kind of selection (Ibanez et al., 2008). Scientists need to be able to apply emerging physical, chemical, and biological approaches to samples (archived or otherwise) in order to understand past and future global trends such as the biological effects of persistent, emerging, and complex mixtures of chemicals that, even at low concentrations, may have an adverse impact, particularly on environments subject to other stresses (Marigomez, 2015). Such information is critical for regulatory bodies, such as the European Chemicals Agency (https://echa.europa.eu), if they are to determine the risk from, and where necessary mitigate against, environmental contamination.

It is likely that a wide range of techniques will increasingly be applied to archived samples, including advances such as pyrosequencing technology (i.e., high-throughput genetic screening) (Koizumi et al., 2009). Archived samples may be used to identify the biological parameters of endangered species (Tanabe and Subramanian, 2011) and track changes in gene frequency. These advances will create innovative functions for ESBs and new information for regulators. However, ESBs may well need to adapt their sampling strategies and specimen-archiving techniques and facilities to enable future use of specimens for novel and emerging analysis and tools. Although there may be long-term preservation of specimens, there is unlikely to be the same long-term preservation of archiving methodologies.

The advent of the information age and "big data" also presents new challenges for ESBs. In terms of information sharing, there are now much better opportunities to publish protocols, SOPs, and methods and to set and adopt common procedures where this is practical. Such harmonization, improving metadata standards (that facilitate data interoperability), and intensive data outputs from new screening techniques will together present significant data challenges. ESBs may need to enhance their computing capabilities to ensure that they can analyze and extract information from extremely large datasets that are generated from the samples they hold.

Finally, at the 2019 ESB conference in Sweden, attendees from all over the globe were informed about new ideas and advances for ESBs. There was discussion over the benefits of incorporating a wider range of sample types in ESBs, including abiotic samples. There are also likely to be a range of new analyses applied to samples, such as biomarkers to identify disease state and increased use of stable-isotope approaches to identify the trophic level of biotic specimens. The latter may itself drive archiving of novel specimen types because they represent specific trophic levels and their inclusion in ESBs would facilitate identification of long-term changes in food web structures. However, adoption of advanced analysis methods, particularly molecular techniques, will require that the ESBs adopt increasingly stringent operational methods to minimize/eliminate cross contamination from air or personal protective equipment. Overall, any growth in the type and number of samples archived will drive adoption of improved ESB management protocols. These could include reducing the required storage volume, for example by storing DNA extracts rather than entire samples, and retaining histopathological samples in the form of high-resolution images. There will also be a likely need to improve routine logistical operations, such as digitizing sample positions within the archive and the volumes/masses of samples retained. This would facilitate faster and easier identification of sample availability and recovery from within the ESB.

In conclusion, what is clear is that at a time when rates of global environmental change are unprecedented, the utility of ESBs, and their ability to look back into the past, has never been more important.

Acknowledgment

J.S. Chaplow and R.F. Shore were supported by the Natural Environment Research Council award number NE/R016429/1 as part of the UK-SCAPE programme delivering National Capability

References

Bond, A.L., Hobson, K.A., Branfireun, B.A., 2015. Rapidly increasing methyl mercury in endangered Ivory Gull (*Pagophila eburnea*) feathers over a 130-year record. Proc. R. Soc. Lond. B Biol. Sci. 282. https://doi.org/10.1098/rspb.2015.0032.

Braune, B.M., Mallory, M.L., Gilchrist, H.G., 2006. Elevated mercury levels in a declining population of ivory gulls in the Canadian Arctic. Mar. Pollut. Bull. 52, 978–982. https://doi.org/10.1016/j.marpolbul.2006.04.013.

Campbell, L.D., 2012. Best practices for repositories. Collection, storage, retrieval, and distribution of biological materials for research. International Society for Biological and Environmental Repositories (ISBER). Biopreserv. Biobank. 10 (2). https://doi.org/10.1089/bio.2012.1022.

Campbell, L.D., Astrin, J.J., DeSouza, Y., Giri, J., Patel, A.A., Rawley-Payne, M., Rush, N., Sieffert, N., 2018. The 2018 revision of the ISBER best practices: summary of changes and the editorial team's development process. Biopreserv. Biobank. 16 (1), 3–6. https://doi.org/10.1089/bio.2018.0001.

Dreyer, A., Neugebauer, F., Rüdel, H., Klein, R., Lohmann, N., Rauert, C., Koschorreck, J., 2018. Halogenated flame retardants in tree samples applied as bioindicators for atmospheric pollution. Chemosphere 208, 233–240.

DuBay, S.G., Fuldner, C.C., 2017. Bird specimens track 135 years of atmospheric black carbon and environmental policy. Proc. Natl. Acad. Sci. U. S. A. 114, 11321–11326. https://doi.org/10.1073/pnas.1710239114.

Espín, S., García-Fernández, A.J., Herzke, D., Shore, R.F., van Hattum, B., Martínez-López, E., Coeurdassier, M., Eulaers, I., Fritsch, C., Gómez-Ramírez, P., Jaspers, V.L.B., Krone, O., Duke, G., Helander, B., Mateo, R., Movalli, P., Sonne, C., van den Brink, N.W., 2016. Tracking pan-continental trends in environmental contamination using sentinel raptors—what types of samples should we use? Ecotoxicology 25, 777–801. https://doi.org/10.1007/s10646-016-1636-8.

Fliedner, A., Rüdel, H., Jürling, H., Müller, J., Neugebauer, F., Schröter-Kermani, C., 2012. Levels and trends of industrial chemicals (PCBs, PFCs, PBDEs) in archived herring gull eggs from German coastal regions. Environ. Sci. Eur. 24 (7). http://www.enveurope.com/content/pdf/2190-4715-24-7.pdf. (accessed 19.02.2019).

Fliedner, A., Lohmann, N., Rüdel, H., Teubner, D., Wellmitz, J., Koschorreck, J., 2016. Current levels and trends of selected EU water framework directive priority substances in freshwater fish from the German environmental specimen bank. Environ. Pollut. 216, 866–876.

Gies, A., Schroeter-Kermani, C., Ruedel, H., Paulus, M., Wiesmueller, G.A., 2007. Frozen environmental history: the German environmental specimen bank. Organohalogen Compd. 69, 504–507. http://dioxin20xx.org/wp-content/uploads/pdfs/2007/07-694.pdf. (accessed 19.02.2019).

Gilchrist, H.G., Mallory, M.L., 2005. Declines in abundance and distribution of the Ivory Gull (*Pagophila eburnea*) in Arctic Canada. Biol. Conserv. 121, 303–309. https://doi.org/10.1016/j.biocon.2004.04.021.

Gómez-Ramírez, P., Shore, R.F., van den Brink, N.W., van Hattum, B., Bustnes, J.O., Duke, G., Fritsch, C., García-Fernández, A.J., Helander, B.O., Jaspers, V., Krone, O., Martínez-López, E., Mateo, R., Movalli, P., Sonne, C., 2014. An overview of existing raptor contaminant monitoring activities in Europe. Environ. Int. 67, 12–21. https://doi.org/10.1016/j.envint.2014.02.004.

Green, R.E., Scharlemann, J.P.W., 2003. Egg and skin collections as a resource for long-term ecological studies. Bull. Br. Ornithol. Club 123A, 165–176.

Ibanez, M., Sancho, J.V., Hernández, F., McMillan, D., Rao, R., 2008. Rapid non-target screening of organic pollutants in water by ultraperformance liquid chromatography coupled to time-of-light mass spectrometry. TrAC Trends Anal. Chem. 27, 481–489.

ISAD(G), 2000. International Council on Archives: General International Standard Archival Description. https://www.ica.org/sites/default/files/CBPS_2000_Guidelines_ISAD%28G%29_Second-edition_EN.pdf. (accessed 25.07.2019).

Klein, R., Paulus, M., Tarricone, K., Teubner, D., 2018. German Environmental Specimen Bank: Guideline for Sampling and Sample Treatment—Bream (*Abramis brama*). Version 2.0.4. https://www.umweltprobenbank.de/upb_static/fck/download/SOP_ESB_Bream_V2.0.4_2018_en.pdf. (accessed 19.02.2019).

Koizumi, A., Harada, K., Inoue, K., Hitomi, T., Yang, H.-R., Moon, C.S., Wang, P., Hung, N.N., Watanabe, T., Shimbo, Ikeda, M., 2009. Past, present, and future of environmental specimen banks. Environ. Health Prev. Med. 14 (6), 307–318. https://doi.org/10.1007/s12199-009-0101-1.

Kotthoff, M., Rüdel, H., Jürling, H., Severin, K., Hennecke, S., Friesen, A., Koschorreck, J., 2019. First evidence of anticoagulant rodenticides in fish and suspended particulate matter: spatial and temporal distribution in German freshwater aquatic systems. Environ. Sci. Pollut. Res. 26 (8), 7315–7325. https://doi.org/10.1007/s11356-018-1385-8.

Leslie, H.A., Leonard, P.E.G., Shore, R.F., Walker, L.A., Bersuder, P.R.C., Morris, S., Allchin, C.R., de Boer, J., 2011. Decabromodiphenylether and hexabromocyclododecane in wild birds from the United Kingdom, Sweden and the Netherlands: screening and time trends. Chemosphere 82, 88–95. https://doi.org/10.1016/j.chemosphere.2010.09.073.

Luo, Y.X., 2018. A review on the historical development of environmental specimen banks. In: E3S Web of Conferences 38., https://doi.org/10.1051/e3sconf/20183801016.

Marigomez, I., 2015. Environmental Specimen Banks International Conference Proceedings. http://www.icesb2015.org/media/upload/Booklet_ICESB2015.pdf. (accessed 24.07.19).

Marigomez, I., Lekube, X., 2017. Environmental Specimen Banks: Sample and Data Inventory in Marine Areas of Arctic and Sub-Arctic Regions. https://ec.europa.eu/research/participants/documents/downloadPublic?documentIds=080166e5b4c5b89d&appId=PPGMS. (accessed 24.07.19).

Movalli, P., Duke, G., Ramello, G., Dekker, R., Vrezec, A., Shore, R.F., Garcia-Fernandez, A., Wernham, C., Krone, O., Alygizakis, N., Badry, A., Barbagli, F., Biesmeijer, K., Boano, G., Bond, A.L., Choresh, Y., Christensen, J.B., Cincinelli, A., Danielsson, S., Dias, A., Dietz, R., Eens, M., Espín, S., Eulaers, I., Frahnert, S., Fuiz, T.I., Gkotsis, G., Glowacka, N., Gomez-Ramirez, P., Grotti, M., Guiraud, M., Hosner, P., Johansson, U., Jaspers, V.L. B., Kamminga, P., Koschorreck, J., Knopf, B., Kubin, E., LoBrutto, S., Lourenco, R., Martellini, T., Martinez-Lopez, E., Mateo, R., Nika, M.-C., Nikolopoulou, V., Osborn, D., Pauwels, O., Pavia, M., Pereira, M.G., Ruedel, H., Sanchez-Virosta, P., Slobodnik, J., Sonne, C., Thomaidis, N., Toepfer, T., Treu, G., Vainola, R., Valkama, J., van der Mije, S., Vangeluwe, D., Warren, B.H., Woog, F., 2019. Progress on bringing together raptor collections in Europe for contaminant research and monitoring in relation to chemicals regulation. Environ. Sci. Pollut. Res. 26, 20132–20136. https://doi.org/10.1007/s11356-019-05340-6.

Newton, I., 1986. The Sparrowhawk. Calton, T. & A.D. Poyser. 396pp.

Newton, I., Wyllie, I., 1992. Recovery of a sparrowhawk population in relation to declining pesticide contamination. J. Appl. Ecol. 29, 476–484.

Poulsen, M.E., Pritzl, G., 1993. The Danish environmental specimen bank—status of establishment. Sci. Total Environ. 139/140, 61–68. https://doi.org/10.1016/0048-9697(93)90008-T.

Prevedouros, K., Jones, K.C., Sweetman, A.J., 2004. European-scale modeling of concentrations and distribution of polybrominated diphenyl ethers in the pentabromodiphenyl ether product. Environ. Sci. Technol. 38, 5993–6001.

Ratcliffe, D.A., 1980. The Peregrine Falcon. Calton, T. & A.D. Poyser. 416pp.

Roselaar, K., 2003. An inventory of major European bird collections. Bull. Br. Ornithol. Club 123A, 253–337.

Rüdel, H., Uhlig, S., Weingärtner, M., 2008. German Environmental Specimen Bank: Guidelines for Sampling and Sample Processing - Pulverization and Homogenization of Environmental Samples by Cryomilling. https://www.umweltprobenbank.de/upb_static/fck/download/IME_SOP_preparation_Dez2008_V200.pdf. (accessed 19.02.2019).

Rüdel, H., Kösters, J., Schörmann, J., 2011a. German Environmental Specimen Bank: Guidelines for Chemical Analysis: Determination of the Elemental Content of Environmental Samples Using ICP-MS. https://www.umweltprobenbank.de/upb_static/fck/download/SOP_ICP-MS_en.pdf. (accessed 19.02.2019).

Rüdel, H., Müller, J., Jürling, H., Bartel-Steinbach, M., Koschorreck, J., 2011b. Survey of patterns, levels, and trends of perfluorinated compounds in aquatic organisms and bird eggs from representative German ecosystems. Environ. Sci. Pollut. Res. 18, 1457–1470.

Tanabe, S., Subramanian, A., 2011. Need for environmental specimen banks in developing countries. Mar. Pollut. Bull. 62 (9), 1933–1934. https://doi.org/10.1016/j.marpolbul.2011.03.020.

Togola, A., Fontan, S.L., Lestremau, F., Margoum, C., 2017. Non-target screening and environmental specimen banking: French perspectives. In: International Conference on Environmental Specimen Banks, Bilbao, Spain. https://www.ehu.eus/en/web/esb2017ff.ffhal01578332f. (accessed 24.07.19).

UBA, 2008. German Environmental Specimen Bank. Concept. Status: October 2008. German Federal Environment Agency (UBA), Berlin. https://www.umweltprobenbank.de/upb_static/fck/download/concept_oct_2008_en.pdf. (accessed 19.02.2019).

Vázquez, M.D., Fernández, J.Á., Real, C., Villares, R., Aboal, J.R., Carballeira, A., 2007. Design of an aquatic biomonitoring network for an environmental specimen bank. Sci. Total Environ. 388 (1), 357–371. https://doi.org/10.1016/j.scitotenv.2007.07.051.

Vo, A.T.E., Bank, M.S., Shine, J.P., Edwards, S.V., 2011. Temporal increase in organic mercury in an endangered pelagic seabird assessed by century-old museum specimens. Proc. Natl. Acad. Sci. U. S. A. 108, 7466–7471.

Walker, L.A., Shore, R.F., Turk, A., Pereira, M.G., Best, J., 2008. The predatory bird monitoring scheme (PBMS)—identifying chemical risks to top predators in Britain. Ambio 37, 466–471.

Walker, L.A., Chaplow, J.S., Potter, E.D., Shore, R.F., 2014. PBMS Archive Holdings: A Predatory Bird Monitoring Scheme (PBMS) Report. Centre for Ecology & Hydrology, Lancaster, UK. 36 pp https://wiki.ceh.ac.uk/download/attachments/134414860/PBMS_Archive_Holdings_2012_FINAL.pdf?version=1&modificationDate=1411034879000&api=v2. (accessed 21.08.2019).

Wieczorek, J., Bloom, D., Guralnick, R., Blum, S., Döring, M., Giovanni, R., Robertson, T., Vieglais, D., 2012. Darwin core: an evolving community-developed biodiversity data standard. PLoS One 7 (1). https://doi.org/10.1371/journal.pone.0029715.

Zhao, J., Becker, P.R., Meng, X., 2015. 2013 International conference on environmental specimen banks: securing a strategy to monitor emerging pollutants in the regional and global environment. Environ. Sci. Pollut. Res. 22 (3), 1555–1558. https://doi.org/10.1007/s11356-014-3715-9.

Index

Note: Page numbers followed by *t* indicate tables.

Kolibri 360 GmbH
Ferdinand-Jühlke-Straße 7
99095 Erfurt, Deutschland
produktsicherheit@kolibri360.de